Lectures and Exhibition:

Physics, Primes, Fractals
and Perspective

ii

Lectures and Exhibition:

Physics, Primes, Fractals and Perspective

The Henry Koerner Center for Emeritus Faculty

Yale University, 2004 – 2014

Frank W. K. Firk

Professor Emeritus of Physics

2014

iv

ISBN – 13: 978 - 1499381979
ISBN – 10: 1499381972

Preface

The Henry Koerner Center for Emeritus Faculty at Yale University opened in 2003. It is housed in a magnificently restored building, the John Pierpoint House (1767), on The Green in New Haven, Connecticut. A most generous gift from Joseph Koerner and Lisbeth Rausing in honor of Joseph's father Henry Koerner, provides the funds necessary to run all aspects of the center, including financial support for those retired faculty who wish to continue teaching in Yale College, and those who wish to continue doing research. Important activities at the center involve formal and informal lectures given by Fellows to their colleagues and their guests. The interaction among members of the Fellowship is quite extraordinary; it is not uncommon to have Emeritus Professors of Anthropology and Zoology attend lectures on Milton or Greek literature. I gave six lectures, presented here, over a period of many years. The topics range from the philosophical basis of Einstein's work to an overview of the theory and practice of Perspective. I have included examples of my drawings and paintings that were part of an exhibit that I had at the Center in 2004.

Contents

Cover:

An Exhibition of Drawings, Paintings and Graphics (April, 2004)

UNDERSTANDING EINSTEIN

March 23, 2006

I shall not talk about the technicalities of space, time, and motion, or the Algebra of Tensors, or the Calculus of Differential Equations. I wish to discuss the *Einsteinian Method* of thinking about, and doing Theoretical Physics. I shall stay away from the risky business of psychoanalyzing the man – the true geniuses who have lived remain beyond analysis by mere mortals.

For almost one thousand years, the *Scientific Method* has formed the basis of research and discovery in all fields of science. Galileo was the first to use the method in a consistent and successful way. He carried out controlled experiments to study the properties of everyday motion of objects. He set out to answer the question – "what property of the motion of an object is related to the force causing the motion?" Is it the speed of the object, or what? He found by experiment that force is related to the rate-of-change of speed, or the acceleration, of the object. Also, he addressed the question – "what is the natural state of motion of an object?" The renowned Aristotle had concluded that the natural state is one of rest; he had based his argument on observation. Galileo realized that the effects of friction and air resistance on the motion of an object are not germane to the question of "pure" motion, he therefore concluded that, in the absence of external forces, an object will continue to move in a straight line, forever.

The immortal Newton followed the *Scientific Method*, up to a point. He did experiments, analyzed data, found trends and correlations in the results, and expressed his findings in mathematical terms. He was one of the two or three greatest mathematicians of all time. He used the results of scientific observations as data for mathematical analysis. He was a masterful Applied Mathematician; his invention of Differential Calculus came about as a result of his need to understand the notion of instantaneous speed. He knew that speed is simply distance traveled/time taken; Newton was interested in this ratio as the time interval tends to zero. He took a great leap, beyond everyday experience when he surmised that the force of the earth causing an apple to fall to the ground is the same force that keeps the moon in "her orb" about the earth, and the same force that keeps the planets in orbit about the sun. Using the universal truth of mathematical equations, he proposed "the universal force of gravity".

This brings us to Einstein. Let me read some key passages from the Herbert Spencer Lecture that he gave at Oxford, June 10, 1933. The lecture was entitled *"On the Method of Theoretical Physics"*. He began, "If you want to find out anything from the theoretical physicists about the methods they use, I advise you to stick closely to one principle: don't listen to their words, fix your attention on their deeds. To him who is a discoverer in this field, the products of his imagination appear so necessary and natural that he regards them, and would like to have them regarded by others, not as creations of thought but as given realities.
These words sound like an invitation to you to walk out of this lecture. You will say to yourselves, the fellow's a working physicist himself and ought therefore to leave all questions of the structure of theoretical science to the epistemologists".

He continued, "Let us now cast an eye over the development of the theoretical system, paying special attention to the relations between the content of the theory and the totality of empirical fact. We are concerned with the eternal antithesis between the two inseparable components of our knowledge, the empirical and the rational, in our development....

We reverence ancient Greece as the cradle of western science. Here for the first time, the world witnessed the miracle of a logical system which proceeded from step to step with such precision that every single one of its propositions was absolutely indubitable – I refer to Euclid's geometry. This admirable triumph of reasoning gave the human intellect the necessary confidence in itself for its subsequent achievements. If Euclid failed to kindle your youthful enthusiasm, then you were not born to be a scientific thinker.

But before mankind could be ripe for a science which takes in the whole of reality, a second fundamental truth was needed, which only became common property among the philosophers with the advent of Kepler and Galileo. Pure logical thinking cannot yield us any knowledge of the empirical world; all knowledge of reality starts from experience and ends with it. Propositions arrived at by purely logical means are completely empty as regards reality. Because Galileo saw this, and particularly, because he drummed it into the scientific world, he is the father of modern physics – indeed, of modern science altogether".

Einstein continued: "If, then, experience is the alpha and the omega of all our knowledge of reality, what is the function of pure reason in science? A complete system of theoretical physics is made up of concepts, fundamental laws which are supposed to be valid for those concepts, and conclusions to be reached by logical deduction. It is these conclusions which must correspond with our separate experiences; in any theoretical treatise their logical deduction occupies almost the whole book....if one regards Euclidean geometry as the science of the possible mutual relations of practically rigid bodies without abstracting from its original empirical content, the logical homogeneity of geometry and theoretical physics becomes complete. We thus assign to pure reason and experience their places in a theoretical system of physics. The structure of the system is the work of reason; the empirical contents and their mutual relations must find their representations in the conclusions of the theory. In the possibility of such a representation lie the sole value and justification of the whole system, and especially of the concepts and fundamental principles which underlie it. Apart from that, these latter are free inventions of the human intellect, which cannot be justified either by the nature of that intellect or in any other fashion *a priori*. These fundamental concepts and postulates, which cannot be further reduced logically, form the essential part of a theory, which reason cannot touch. It is the grand object of all theory to make these irreducible elements as simple and as few in number as possible, without having to renounce the adequate representation of any empirical content whatsoever.

The view I have just outlined of the purely fictitious character of the fundamentals of scientific theory was by no means the prevailing one in the eighteenth and nineteenth centuries...."

Einstein then considers the Newtonian method; he continued: "Newton, the first creator of a comprehensive, workable system of theoretical physics, still believed that the basic concepts and laws of his system could be derived from experience....Actually, the concepts of time and space appeared at the time to present no difficulties. The concepts of mass, inertia, and force, and the laws connecting them, seemed to be drawn directly from experience. Once this basis is accepted, the expression for the force of gravitation appears derivable from experience, and it was reasonable to expect the same in regard to the other forces.

We can see from Newton's formulation of it that the concept of absolute space, which comprised that of absolute rest made him feel uncomfortable; he realized that there seemed to be nothing in experience corresponding to this last concept. He was not comfortable about the introduction of forces acting at a distance. But the tremendous practical success of his doctrines may well have prevented him and the physicists of the eighteenth and nineteenth centuries from recognizing the fictitious character of the foundations of his system.

The natural philosophers of those days were, on the contrary, most of them possessed with the idea that the fundamental concepts and postulates of physics were not in the logical sense free inventions of the human mind but could be deduced from experience by abstraction – that is to say, by logical means. A clear recognition of the erroneousness of this notion really only came about with the General Theory of Relativity (Einstein's crowning achievement) that showed one could take account of a wider range of empirical facts, and that, too, in a more satisfactory and complete manner, on a foundation quite different from the Newtonian. But quite apart from the question of the superiority of one or the other, the fictitious character of fundamental principles is perfectly evident from the fact that we can point to two essentially different principles, both of which correspond with experience to a large extent; this proves at the same time that every attempt at a logical deduction of the basic concepts and postulates from elementary experiences is doomed to failure.

If then, it is true that the axiomatic basis of theoretical physics cannot be extracted from experience but must be freely invented, can we ever hope to find the right way? Nay, I say more, has this right way any existence outside our illusions? Can we hope to be guided safely by experience at all when there exist theories (such as classical {Newtonian} mechanics) which to a large extent do justice to experience, without getting to the root of the matter? I answer without hesitation that there is, in my opinion, a right way, and that we are capable of finding it....

......*Experience remains, of course, the sole criterion of the physical utility of a mathematical construction. But the creative principle resides in mathematics. In a certain sense, therefore, I hold it true that pure thought can grasp reality, as the ancients dreamed".* Einstein was a modern Pythagorean.

Many of the most important advances in theoretical physics in the 20th-century have been strongly influenced by the Einsteinian method. Let me give you one example. In 1923/24 Prince Louis deBroglie proposed one of the most profound concepts in the history of physics namely particle-wave duality – a fundamental concept that had no experimental basis at the time. deBroglie's line of pure thought was the following:

The basic concept underlying the dynamics of all matter and waves is the event – a description of when and where something happens. In its most familiar form, we label it by a symbol, t, for the time, and by three symbols x, y, and z for the position. These numbers are given with respect to arbitrarily chosen origins. Einstein's theory of Special Relativity (no gravity present) centers around quantities we call *invariants*. An example of an invariant is Pythagoras' theorem in geometry, namely: $x^2 + y^2 = $ (length of hypotenuse)$^2 = r^2$ — an *invariant under rotations and translations in space.* (If we rotate our axes so that we study the same length r, and our new coordinates are x' and y', then $x'^2 + y'^2 = r^2$). This is the only mathematical form that is invariant under rotations and translations in flat, Euclidean space. (If $x^4 + y^4 = $ invariant, we would always face the same direction in which we were born). If, now, we describe an event by [t, x], we ask the question – is $t^2 + x^2$ an invariant under rotations in space-time? This is in the spirit of the Einsteinian method. It turns out that the answer to our question is "no". What is found, however, is that $t^2 - x^2$ is in invariant! (Here, we are using units in which the speed of light is unity). It is not possible for our limited minds to grasp the notion of the difference between two squares being equal to the interval between two events. But that is what Nature prescribes. Einstein searched for other invariants associated with material particles and found that, if E denotes the total energy of a moving particle, and p denotes its momentum (essentially mass multiplied by velocity) then $E^2 - p^2$ = invariant with respect to different observers moving uniformly in space-time (so-called inertial observers). More than a decade after Einstein's findings, deBroglie searched for possible invariants for wave motion. He found the remarkable form

$\omega^2 - k^2$ = invariant for all inertial observers. Here, ω = $2\pi\upsilon$ where υ is the frequency of the wave, and k = $2\pi/\lambda$ where λ is the wavelength. deBroglie went further and argued that if $E^2 - p^2$ is an invariant that stems from an event [t, x], and $\omega^2 - k^2$ is an invariant that stems from the event, then E is logically connected to ω, and p is logically connected to k. He surmised that the quantities are proportional to each other - the simplest possible connection. To agree with Planck's discovery for the quantum of radiation (waves), announced in December, 1900, deBroglie stated that the constant of proportionality must be Planck's constant, h (divided by 2π). deBroglie therefore proposed, *in complete generality*, that

$$E = h\omega/2\pi, \text{ and } p = hk/2\pi,$$

and therefore,

$$E = h\upsilon \text{ and } p = h/\lambda.$$

These equations link, forever, our notion of particles and waves. The extremely small value of Planck's constant means that we do not experience the "wave" properties of particles in our everyday lives. However, on the atomic scale, we observe both the particle-like and the wave-like properties of matter. The observation of particle-wave duality was not made until three years after deBroglie's theory had been published. His revolutionary idea is a classic example of *pure reason* at work in the scientific enterprise.

The Physics of Nuclear Power and Related Issues

March 23, 2009

Abstract. An overview of the fundamental physical principles involved in the generation of electric power from nuclear reactors is given. The key question addressed is: why is the generation of electric power using nuclear reactors an important alternative to traditional methods that derive energy from burning fossil fuels? The answer to this question requires an understanding of the basic nuclear physics involved. The economics of nuclear power, and the environmental issues that are important in formulating our national energy policy, are discussed.

At the present time, it is estimated that 87% of primary energy production throughout the world comes from burning fossil fuels. The remaining sources are hydroelectric (6%), nuclear power (6%), and all other sources (geothermal, solar, wind, wood, and waste) (1%). These percentages are averages, and they do not reflect the major differences in energy use that exist between the highly–industrialized countries and the under–developed countries. The sources of global energy use are illustrated in the following chart.

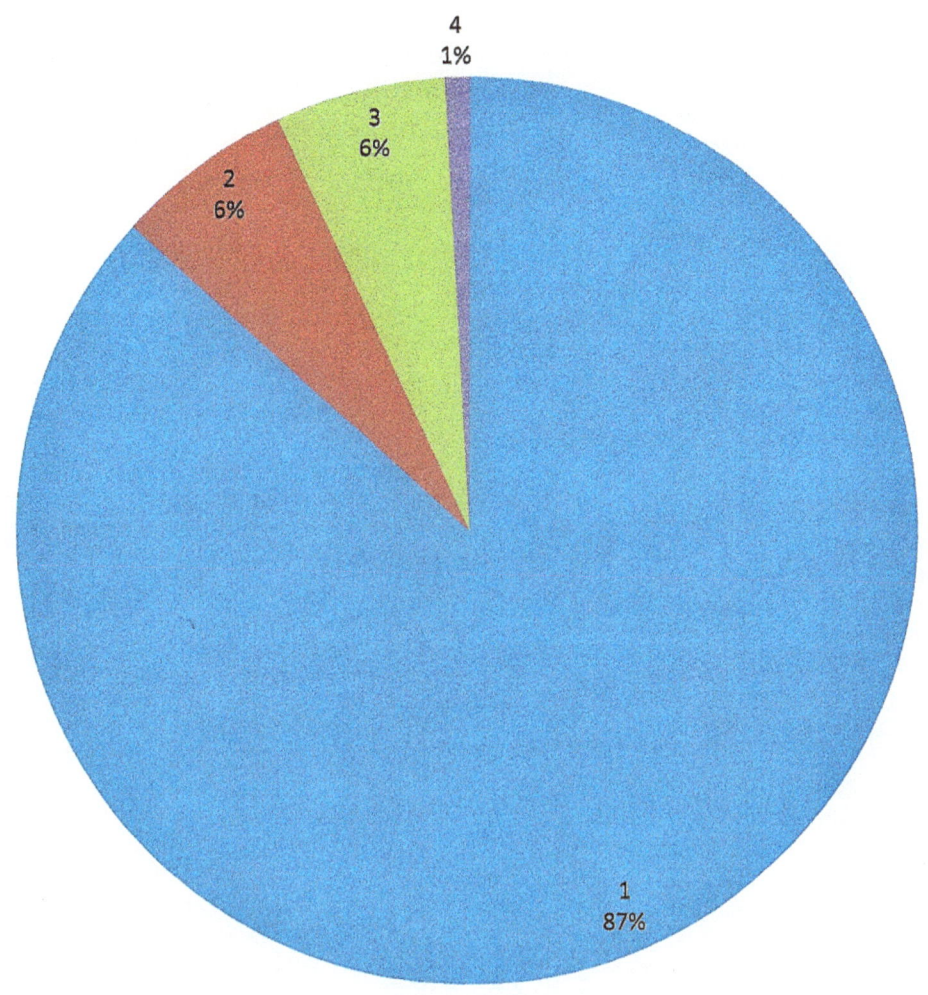

1. Burning fossil fuels: 87%
2. Hydroelectric power: 6%
3. Nuclear power: 6%
4. Other (geothermal, solar, wind, wood, and waste): 1%

The fossil fuels took millions of years to form; they are non-renewable and they will be completely used up in the next few hundred years. The rate of depletion of the estimated world reserves of coal is shown in the following diagram. Here, the set of model parameters is based on the current reserves and consumption rate, and on an annual increase in the rate of 2%.

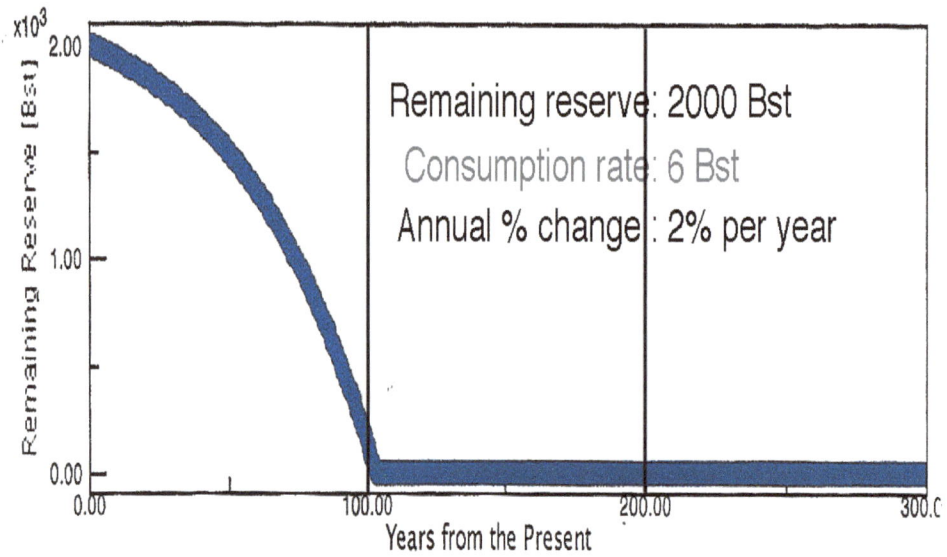

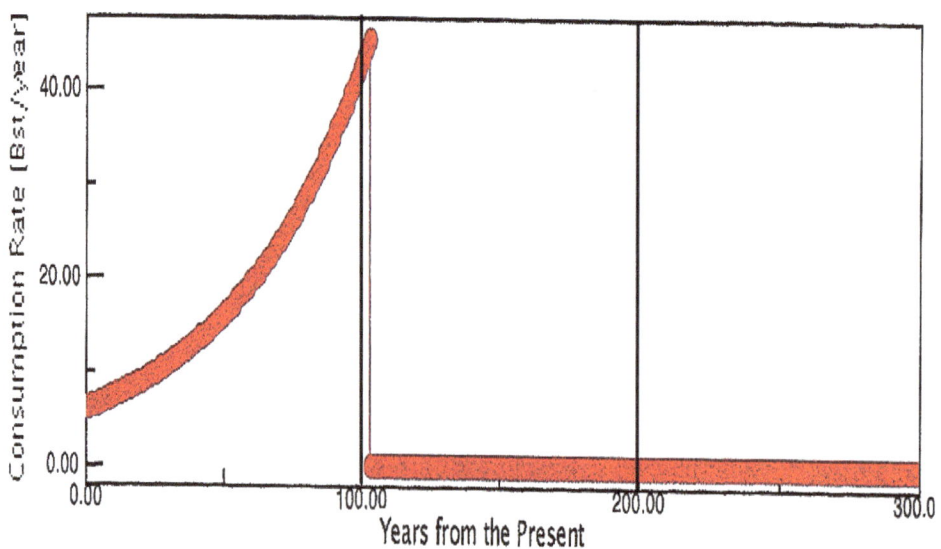

(The unit Bst is 1 billion short tons)

The global emission of carbon from each of the major fossil fuel types is shown in the next diagram:

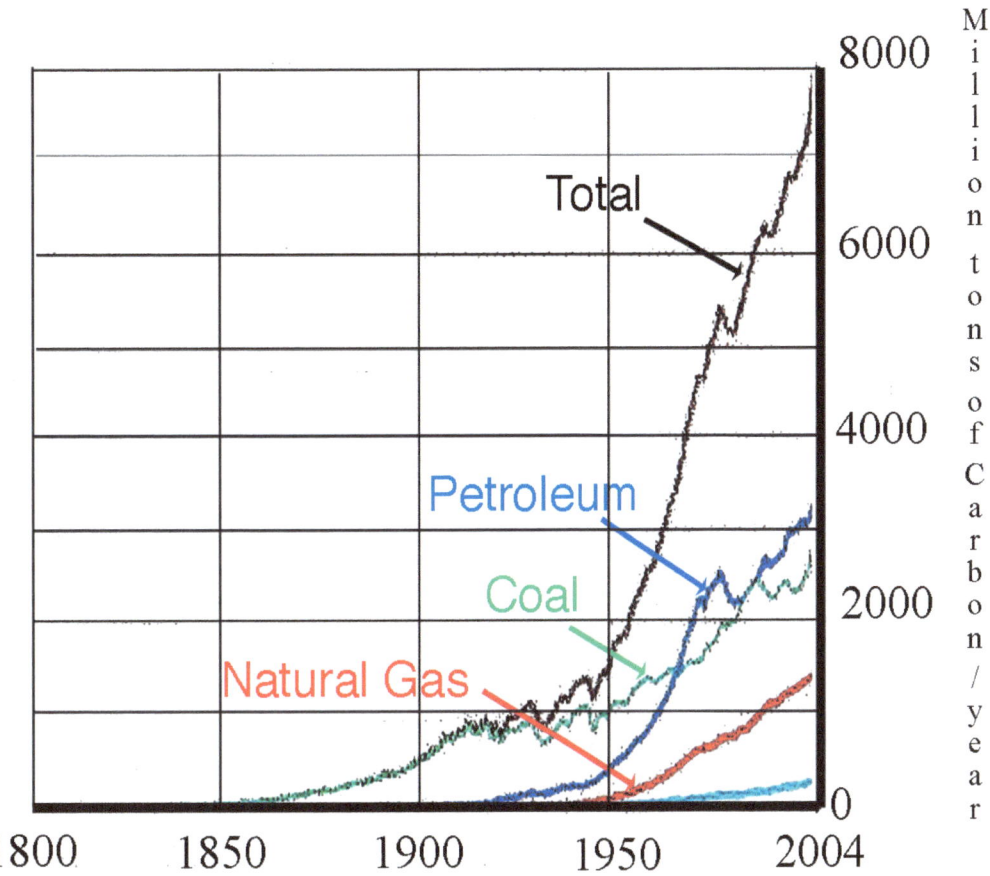

Global fossil carbon emission by fuel type
1800 – 2004
(1 ton of carbon produces 3.7 tons of carbon dioxide)

Currently, natural resources can absorb about 50% of the amount of carbon dioxide formed. Approximately 15 billion tons of CO_2 per year therefore pollute our atmosphere. This is a sobering fact. The need for alternative, non‑polluting power sources is clear. I shall discuss one of them namely, *nuclear power*.

If we study the structure of matter at the atomic level, we find that an atom consists of a central, relatively massive object called the nucleus, surrounded by a cloud of electrons. The radii of atoms are typically 10^{-9} meters, and the radii of nuclei are typically 10^{-14} meters. Nuclei consist of positively–charged *protons*, each with a radius $\approx 10^{-15}$ meters, and electrically–neutral *neutrons*, each with a radius equal to that of the proton. These "nucleons" are held together with a very strong force that has a very short range ($\sim 10^{-15}$ meters). We have no direct experience of the nuclear force. In addition to the strong nuclear force that exists between proton-proton, neutron-neutron, and neutron-proton pairs, there is a long-range electric force between the positively charged protons. The balance between the short-range, *attractive* nuclear force between pairs of nucleons, and the long-range, *repulsive* electric force that acts on all protons, is a delicate one. This balance determines the limiting size and stability of nuclei.

Complex "light nuclei" are built-up by adding neutrons and protons to a given system. For example,

$p^+ \, O$ $\qquad\qquad\qquad$ $p^+ \, OO \, n^\circ$ $\qquad\qquad$ $p^+ p^+ \, OOOO \, n^\circ n^\circ$

nucleus of hydrogen $\quad$ nucleus of deuterium $\quad$ nucleus of helium

and nuclei such as calcium–40 (20 protons and 20 neutrons), written

$^{40}_{20p} \, Ca_{20n}$. The heavy nuclei (mass > 50) are formed by a different mechanism that involves the radiative capture of neutrons. Light nuclei have the interesting property that the number of neutrons is often equal to the number of protons, forming a particularly stable arrangement. This is an important observation that is consistent with the theory of the "charge–independence" of the nuclear force. For heavier nuclei, the repulsive electric force between all the protons makes it more difficult to add equal numbers of protons and neutrons; steadily, the number of neutrons in a nucleus becomes much greater than the number of protons. Eventually, the repulsive electric force is so great that it becomes impossible to increase the size of the very dense nucleus, held together by the extremely short–range nuclear force acting between pairs of nucleons. This limit occurs naturally at uranium with 238 nucleons – 146 neutrons and 92 protons; the periodic table of the natural elements ends at this point. In some nuclei, with masses less than U–238, the balance is so delicate that they spontaneously disintegrate, emitting nuclear particles or radiation; this is the phenomenon of radioactivity.

Every complex nucleus (a "bound state" of nucleons) has a mass that is less than the mass of its "free" constituents. For example, the nucleus of helium (called an alpha–particle) contains 2 neutrons and 2 protons; its mass is measured to be

mass of alpha–particle = 4.00153 atomic mass units (u)

whereas, in the free state

mass of 2 protons = 2. 01456 u

and

mass of 2 neutrons = 2.01732 u

giving a total free mass = 4.03188 u

The difference, Δm = 0.03035 u.

This may not seem to be very large, but that has to do with the unusual units in which the mass is measured.

The important conversion factor is

1 u = 1.66054 x 10^{-27} kilograms

= 931.494 million electron volts/(speed of light)^2

= 931.494 MeV/c^2

(Here we see Einstein's famous E = mc^2)

It is customary to use the "electron–volt" as a unit of energy in Modern Physics: 1 eV is the energy acquired by an electron when it is accelerated through a potential difference of 1 volt. We therefore find that the difference in mass (energy) between the bound state of the alpha–particle and the free state of the four nucleons is

Δm = 0.03035 u $\rightarrow$ ΔE = 28.3 MeV (million electron volts)

The "energy difference" ΔE is called the **nuclear binding energy**. It is a measure of the energy needed to break apart the nucleus into its constituent parts. (Einstein's $E = mc^2$ includes not only the energy of motion (kinetic energy) but also the "potential energy"). How does the nuclear binding energy compare with typical atomic binding energies and chemical binding energies? In the simplest case, we find that the energy needed to take apart (ionize) a hydrogen atom is only 13.6 eV. (The hydrogen atom consists of a central nucleus, a proton, and an orbiting electron). Nuclear binding energies are typically one million times greater than atomic and chemical binding energies, and this fact is of key importance in understanding nuclear power generation.

We now consider a remarkable nuclear phenomenon, discovered in the late 1930's, called nuclear fission. In certain heavy nuclei, near the end of the periodic table, the internal forces are so delicately balanced that nuclear transmutations take place in which a nucleus spontaneously splits into two lighter nuclei, accompanied by the emission of neutrons and high-energy radiation (gamma-rays). The process is known as nuclear fission. The masses of the fission fragments are typically in the ranges 90 to 100 mass units and 130 to 140 mass units. The reason for the asymmetry in the two masses will be discussed later. It was discovered that, in certain nuclei, fission is induced by bombardment with other nuclear projectiles, particularly low-energy neutrons (energies below about 1000 eV). The process is illustrated in the following diagram.

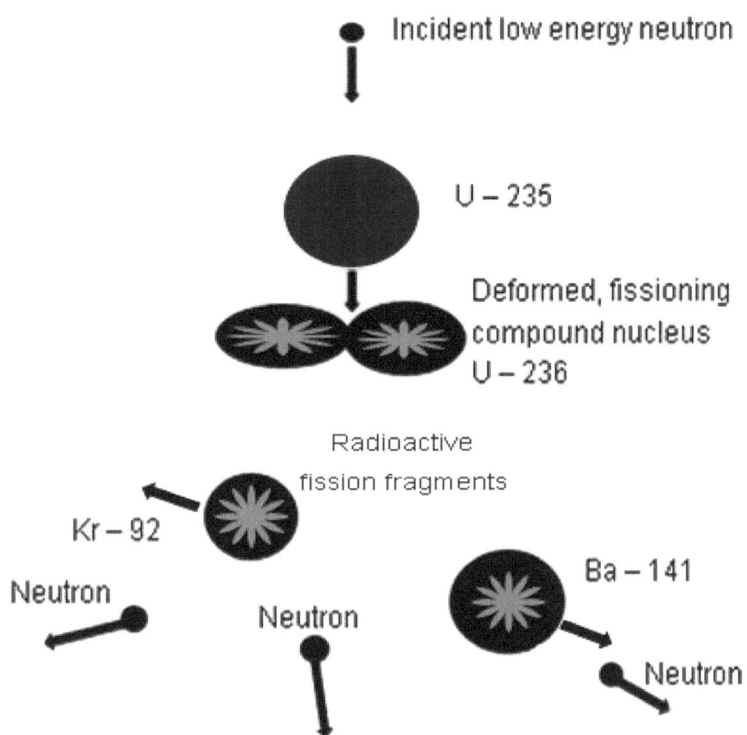

Here, the fission fragments are krypton–92 and barium–141. Heavy nuclei behave like liquid drops; they have remarkably uniform densities and nuclear surface tensions. The addition of energy to most nuclei can create complex nuclear collective motions.

In the most favorable cases (U–235, and Pu–239), the interacting neutrons can generate strong "resonances" that frequently decay with a high probability by fission. At the beginning of the *nuclear power age*, it was important to study these interactions to find out where the resonances occur as a function of the energy of the interacting neutrons. It was also important to study resonances in non–fissile materials that "poison" the true fission process by removing neutrons from the system. I worked in this field of neutron research at the Atomic Energy Research Establishment, Harwell, England from 1952 until 1965. Harwell was the premier government research laboratory in England at that time. It was comparable in size and mission to the Brookhaven National

Laboratory on Long island and the Oak Ridge National Laboratory in Tennessee (where I worked as a visiting scientist in 1960 – 61). Typical neutron resonant structure in the compound nucleus n + ^{238}U (a nucleus that fissions at neutron energies greater than 1 MeV) that we studied in the mid–1950's is shown:

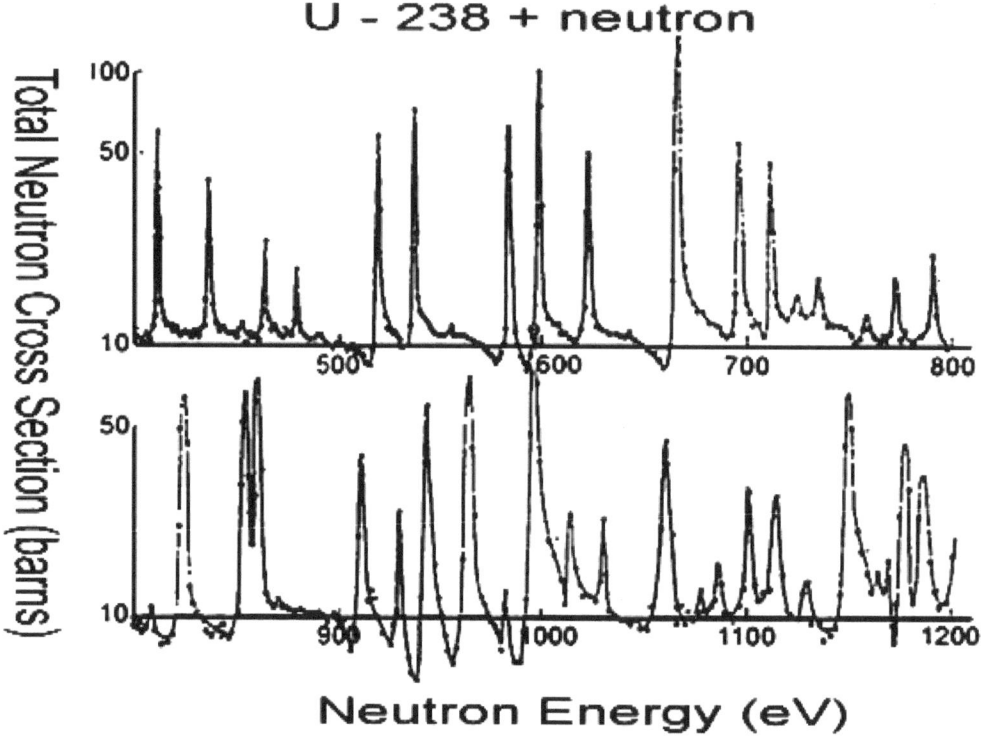

When my colleague Michael Moxon and I began this experiment, *no* resonances had been observed in n + U–238 at these energies. The vertical scale gives the probability of a neutron interaction. The resonances "decay" by neutron or gamma-ray emission (they do not fission at these low energies), and therefore U–238 removes neutrons from the reaction process.

The binding energies of all nuclei have been accurately measured. If the values of the binding energies are normalized by dividing by the number of nucleons in each nucleus we obtain a curve of the *"binding energy per nucleon*" as a function of the mass of the nucleus, shown in the following graph.

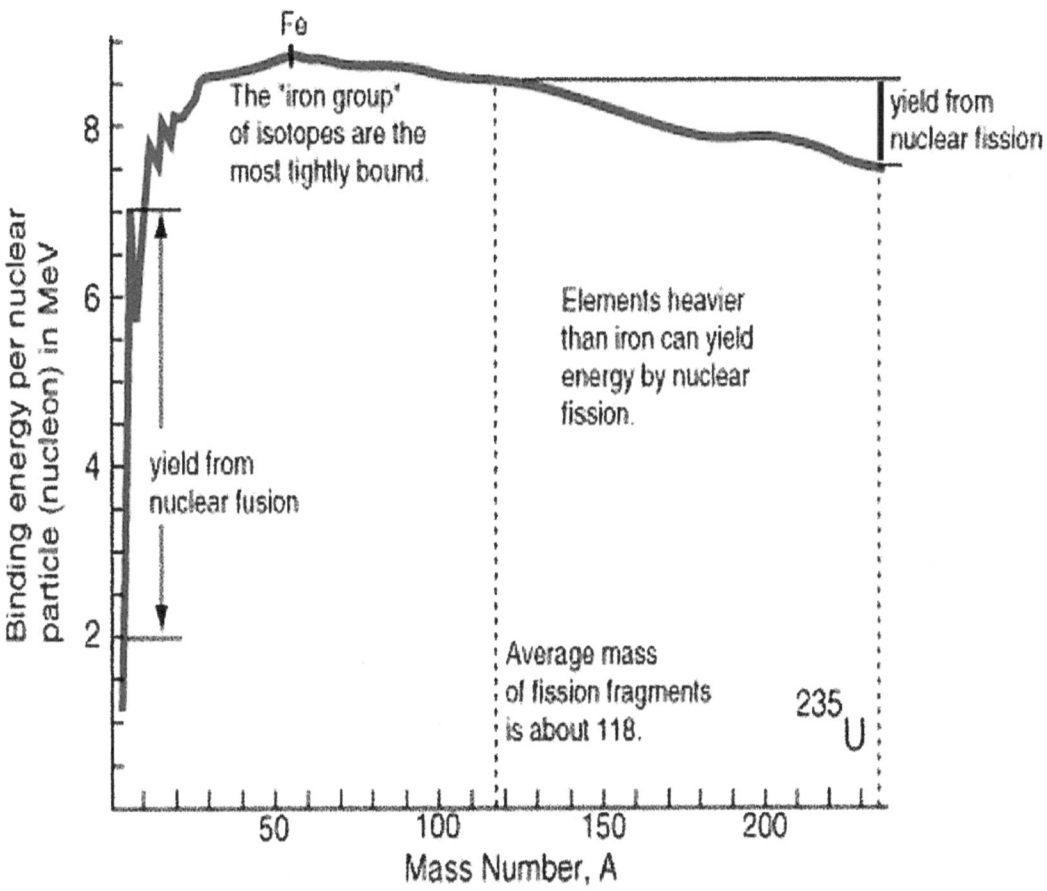

The curve is of fundamental importance not only in understanding nuclear power but also in understanding the formation and relative abundances of the elements in the universe. The slight asymmetry in the shape of the curve about the average mass number of the fission fragments results in the asymmetry of the masses of the two fragments produced in the fission process.

We see that when a nucleus such as U–235 fissions, and becomes two new nuclei, each with mass numbers about 118, a very large amount of energy (almost 200 MeV) is released. The energy is calculated by multiplying 236 (235 + 1 (for the absorbed neutron)) by the difference in binding energy per nucleon that is seen to be between 0.8 and 0.9 MeV. Typically, two or three free neutrons are produced in the fission process together with a burst of high–energy gamma–rays. The total energy released is divided up as follows:

kinetic energy of the two fission fragments	≈ 170 MeV
kinetic energy of the neutrons	≈ 5 MeV
energy generated by all other reaction products	≈ 25 MeV.

The fission fragments have about the same ratio of neutron number, N, to proton number, Z, as the parent nucleus (N/Z $\approx$ 3/2), and therefore they are highly unstable. *These fragments are the sources of the intense radiation associated with the waste produced in nuclear reactors.*

The three (sometimes two) free neutrons that are produced in each fission event are of the greatest importance. Typically, one neutron is absorbed by an impurity nucleus, such as U–238, another escapes from the system without colliding with any nucleus, and the third neutron interacts with U–235 to induce another fission. In this way, it is possible for a self–sustaining "chain reaction" to occur. A typical chain reaction is shown.

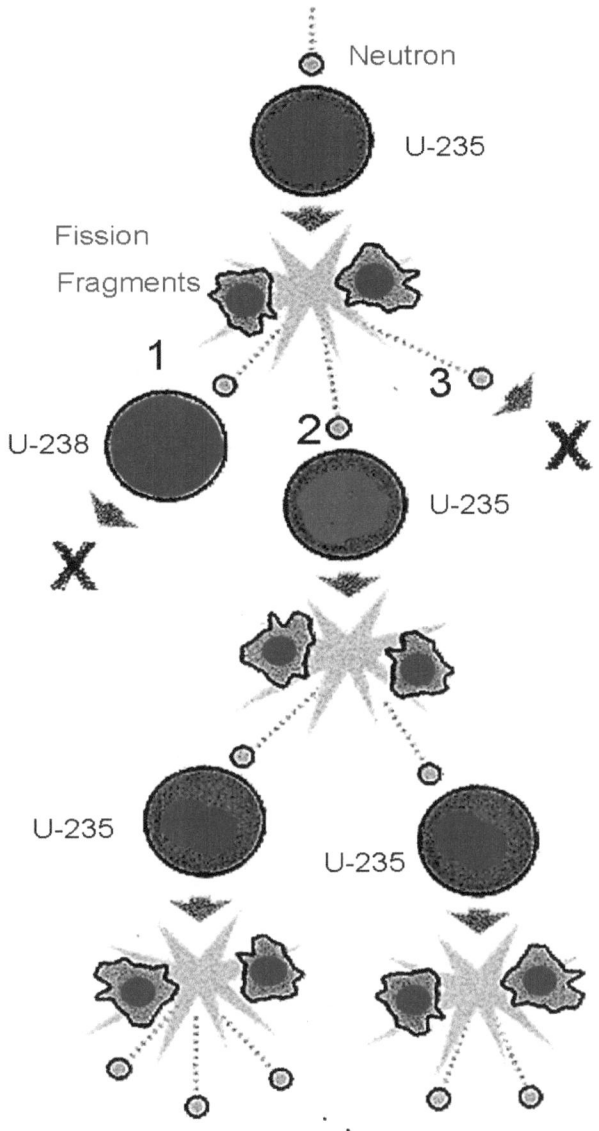

In a nuclear reactor, the rate of the chain reaction is controlled by inserting rods that contain neutron–absorbing material, such as cadmium or boron–10. These rods reduce or increase the number of neutrons available for inducing fissions; they are "fail–safe" rods that automatically shut down the reactor in the event of a problem. In order to achieve a high rate of fission of the U–235 nuclei, it is necessary to surround the fissionable material with a "moderator" that is made of hydrogenous material (typically water or, in some cases, heavy (deuterated) water) with which the neutrons collide. Multiple collisions occur thereby lowering the average fission neutron energy (about 1 MeV) into the "resonance energy region" where the probability of a fission interaction is greatly increased. In natural uranium, 0.7% is in the form U–235 and the rest is U–238. A nuclear reactor requires enriched uranium that contains about 3% of U–235 to sustain a controlled chain reaction in a moderated medium. The low–percentage enrichment needed is a consequence of the fact that the probability of fission in U–235 is a factor of hundred greater than the probability of fission in U–238 at low, moderated neutron energies. An uncontrolled chain reaction, typical of a nuclear bomb, requires the use of pure U–235. The process of separating the natural uranium to obtain such purity is technically challenging, expensive, and time-consuming.

The highly–energetic fission fragments interact with the bulk material in which fissions occur, and they heat it up. The intense heat produced is used to convert water into steam. (The water acts as a moderator and a coolant). The steam then drives turbines to generate electricity, in the standard way. The energy density in the core of a nuclear reactor is very high, and a closed loop of water (often pressurized) is required to cool the core. The loop contains highly–radioactive material and, therefore, it is completely isolated from the outside world. Many nuclear reactors are sited alongside large rivers or lakes, or the sea to provide sources of cooling for the very hot water/steam loop that drives the electric power-generating turbines. A typical nuclear power plant is shown.

A view of the nuclear power plant at Cattenom in France.

Nuclear power plants release no life–threatening radioactivity, no particle pollution, and no greenhouse gases. White plumes from cooling towers are non-radioactive water vapor. In France, the problem of the disposal of radioactive waste has been solved.

Fossil-fuel power plants used to generate electricity contribute to particle pollution, global warming, and to the low-level radioactivity of our surroundings.

A schematic diagram of a pressure-water reactor is shown.

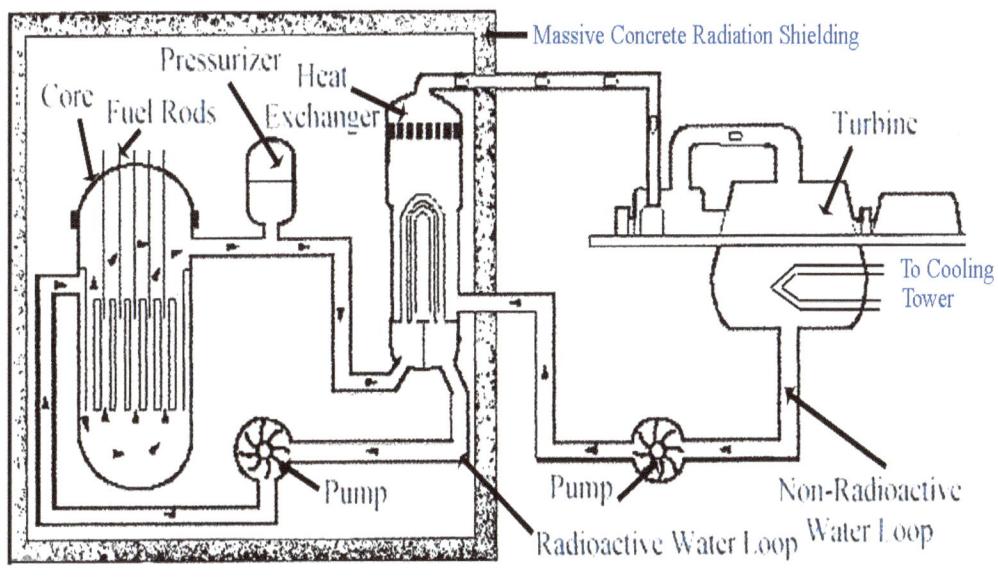

All radioactive elements of the system are contained in the massive concrete radiation shielding. The heat exchanger enables the cooling process to be carried out in a non-radioactive environment. The very hot nuclear waste is stored locally in deep- water tanks where the radioactivity decays over a period of years. Not until the activity has decreased by more than 90% is the waste removed to permanent storage sites where it will remain for many thousands-of-years, under the strictest control. In the United States, an extensive permanent storage facility has been built at Yucca Mountain in Nevada at a cost of more than 10 million dollars. The use of this site as a national facility is not, however, supported by President Obama and his advisers; it is therefore unlikely that it will ever operate. The only large, deep geological repository currently available is near Carlsbad, New Mexico. It is expected to continue operations until 2070.

The only nuclear power plant accident of note in the United States occurred at Three Mile Island, on March 28, 1979. In order to put that accident in perspective, it is necessary to discuss, *quantitatively*, our knowledge about nuclear radiations of all kinds, and the relation between radiation dose and living tissue. We are interested in a form of radiation called *ionizing radiation* that can penetrate cells and generate ions in the contents of the cells. The generic term *ionizing radiation* includes:

x–rays, gamma–rays, electrons, neutrons, and alpha particles.
The units of measurement of radiation dose are

the **rad**: the dose of radiation that causes 0.01 joules of energy
to be absorbed by 1 kilogram of matter.

This unit is no longer in standard use because it does not take into account the known effects of different types of ionizing radiation.

the **rem**: the energy transferred to a cell depends on the type of ionizing radiation. To take into account the differences, a "quality factor", Q, is introduced in the definition of a new unit called the rem ("roentgen–equivalent–man"):

rem = rad • Q

where Q ≈ 1 for x–rays and gamma–rays, Q ≈ 5 for neutrons, and alpha–particles have Q ≈ 20.

(The International Commission on Radiation Units and Measurements recommends the use of the gray (Gy, where 1Gy = 100 rad) in place of the rad, and the sievert (Sv, where 1Sv = 100 rem) in place of the rem. These units are used world wide, except in the United States). In the US, radiation doses are generally reported in units of millirem = mrem = 0.001 rem.

The following list gives typical radiation doses in mrem

Lethal whole–body dose in a short period	500,000
Radiation sickness (if absorbed in a short period)	100,000
Maximum permitted annual dose to public (above background)	170
Annual natural background in Boston (without Radon)	102
Annual natural background in Denver (without Radon)	180
Dose to colon in barium enema	1,500
Dose in mammagram	300
Dose from single full–body CT scan	4,500
Annual person–dose in US from fallout (weapons + Chernobyl)	0.06
Average dose/person within10 miles of Three Mile Is. (> 3/28/79)	8
Average dose /person living for 1 year at fence of nuclear plant	< 0.7

The accident at Chernobyl in the former Soviet Union, generated a dose to those living close to the power plant greater than 40,000 mrem. More than 100 plant workers and fire fighters at the Chernobyl power plant received high dose rates between 70,000 and 1,340,000 mrem, and suffered serious radiation sickness; 28 died from the extreme exposure. At such high radiation levels, the ionizing radiation breaks down molecules, directly, and produces very reactive free radicals that attack local cells. The molecular disruption exceeds the capacity of the body to repair the damage, and causes mutations in replicating cells. The long–term effects of the Chernobyl accident have been studied in detail. Approximately 1,800 cases of thyroid cancer have been reported in the seriously contaminated areas, mostly in children.

The design of the Chernobyl reactor was seriously flawed, and the construction and operation of the plant did not meet even the lowest levels of safety required anywhere else in the world. This one–of–a–kind event damaged the credibility of the nuclear power industry more than all other accidents, combined. Since that time, the design, construction and operation of the latest nuclear power plants in various countries, have improved greatly; the probabilities of risks of even minor accidents have been reduced to very low levels.

It is of interest to study our *annual* natural exposure to background radiation, we find

cosmic radiation: 27 mrem; the value increases with altitude, for example, people living in Denver, Co receive an annual radiation dose of about 50 mrem.

rocks and soil: 28 mrem; a value that depends on the local geology. Those living in Louisiana receive about 15 mrem whereas those living on the Colorado plateau are subjected to 140 mrem.

body radioactivity: 40 mrem; this dose comes, mostly, from the natural radioactivity of potassium–40 within our bodies.

It is clear from the extensive studies of nuclear radiation, and its effect on living cells, that the risks of damage to humans from well-designed and operated nuclear power plants is much less than the risks from natural, and other man–made sources.

The initial costs of constructing and commissioning a modern nuclear power plant are high when compared with the start–up costs of fossil fuel plants. The high costs are associated with the need for very high quality, often rare, construction materials, mining and enrichment of uranium, training of nuclear engineers (now a rare breed following the Three Mile Island accident), training of expert operators, and with the handling and disposal of hazardous radioactive waste. However, long-term economic benefits accrue because of the absence of atmospheric particle pollution and greenhouse gas emission by nuclear power plants. The ensuing benefits to the environment, both locally and globally, are great.

The known reserves of natural uranium will provide the necessary nuclear fuel for power reactors for the next several hundred years – a limited lifetime comparable with that of the fossil fuels. This important fact was fully appreciated from the very beginning of the nuclear power age. Prototype reactors, known as "breeder reactors", were therefore developed. The first one that operated successfully was at the National Reactor Testing Station in Idaho in 1951. Breeder reactors generate new fissionable material (either Pu–239 or U–233) at a greater rate than they consume such material. Research programs in this field took place in the US, UK, and the USSR throughout the 1950's. Later, the French and Japanese governments supported extensive development programs that resulted in successful breeder reactors. The cost of these breeder reactors was not commercially viable at the time, and the national programs in all countries continued along conventional lines. Breeder rectors require elaborate reprocessing plants; at this time, none exist in the United States. In breeder reactors that generate large quantities of toxic plutonium, extreme safeguards must be put in place to prevent this weapon–making fuel from falling into the wrong hands.

A conventional reactor consumes less than 1% of the natural uranium required in a fuel cycle whereas a breeder reactor can, with appropriate reprocessing, use almost all the initial fissionable material. Furthermore, breeder reactors can be designed to use Th–232 as the initial fuel – globally, thorium is much more abundant than uranium. This fact has led India to engage in research programs that use thorium in place of uranium as the primary fuel; in India, approximately 30% of all thorium deposits are known to exist whereas less than 1% of all uranium deposits are to be found there.

Our new President has stated that he is in favor of building more than forty new nuclear power plants in the next decade. This ambitious plan presents us with serious technical and economic challenges. The science and engineering of nuclear reactors is well–understood, and the external radiation associated with them can be reduced to levels that do not pose a threat to the health of all people. However, it is essential that we implement a long–term national plan that deals with two outstanding problems namely, the handling of toxic waste, and the security of weapons–grade nuclear fuel. In developing such a plan, we would do well to learn from the French who have demonstrated the great benefits that come from a plan that has been drawn up by scientists, economists, industrialists and politicians, with the support of the population at large.

The Mysterious Primes

October 13, 2010

Abstract

"God may not play dice with the universe, but something strange is going on with the prime numbers"

Paul Erdos (1913 – 1996), distinguished mathematician.

The natural numbers 1, 2, 3, . . . are an essential part of our everyday lives. Nonetheless, the theory of numbers, even at the most basic level, is not part of the mathematics curriculum in most schools and universities. Although the prime numbers 2, 3, 5, 7, 11, 13, . . . play a key role in Number Theory, their properties remain elusive in spite of intense study over the centuries. If, for example, we find that the number 997 is prime, we have no formula that will give the next prime, 1009. In this lecture, some of the few established properties of the primes, and many unresolved, intriguing questions associated with them, will be discussed.

A **prime number** is any natural number (positive integer) greater than 1 that cannot be expressed as the product of two smaller natural numbers. (Its only divisors are 1 and the number itself).

The first few are 2, 3, 5, 7, 11, 13, 17, 19, 23, 29, 31, 37, 41, 43, · · ·

If n ≥ 2 is not prime it is said to be **composite**.

The number 1 is neither prime nor composite; it is called a **unit**.

The Fundamental Theorem of Arithmetic:

Every natural number greater than 1 can be expressed as the product of one or more primes. This product is unique, up to rearrangement.

For example, we find

$$99 = 3^2 \bullet 11 = 11 \bullet 3^2; \ 100 = 2^2 \bullet 5^2; \ 104 = 2^3 \bullet 13.$$

It is because of the fundamental theorem that 1 is not considered to be prime; for example, if 1 were prime then $1^{100} \bullet 3^2 \bullet 11$ and $3^2 \bullet 11$ would be different factorizations of 99, and this would violate the uniqueness property of the fundamental theorem.

The primes are the building blocks of natural numbers with respect to multiplication.

Euclid's Theorem:

There are infinitely many primes.

Proof by contradiction:

Suppose that there is a finite number, n, of primes labeled $p_1, p_2, p_3, \cdot\cdot\ p_n$ (for example only 2, 3, 5, 7)

Multiply the primes together and add 1:

$P = p_1 \bullet p_2 \bullet p_3 \bullet \ldots p_n$ and add 1 to give $P + 1$.

$(2 \bullet 3 \bullet 5 \bullet 7 + 1 = 211)$

$P + 1$ is an integer greater than 1; it is either a new prime not contained in the list $p_1 \ldots p_n$, or it is composite. If composite, it must be divisible by some prime. It cannot, however, be divisible by any prime in the list because each one gives a remainder 1. Therefore, $P + 1$ is either prime or a composite divisible by a prime not in the list $p_1 \ldots p_n$. In either case, the list is incomplete, proving that there must be infinitely many primes. We do not know how to factor very large numbers – the basis of many modern cryptographic methods. (At this time, it is not possible to factor 200+ digits in a lifetime!).

Finding the Primes

More than two thousand years ago, the Greek mathematician Eratosthanes understood the importance of prime numbers. He devised a method to calculate them, known as *sieving*; it remains the basis of all methods currently used to calculate primes. The following example illustrates the method in its simplest form. Construct a 20 x 20 array of consecutive numbers, omitting 1. Delete all the numbers divisible by 2 except 2. Continue by deleting all the numbers divisible by 3, except 3; continue in this way until the final divisor of 19 has been reached. The remaining numbers are 2, 3, 5, 7, 11, 13, 17, 19. ... 383, 389, and 397. These are all the prime numbers below 400. (Note that 19 is the largest prime divisor less than √400 = 20).

	2	3		5		7			11	13				17	19
	2	3		5		7			11	13				17	19
		23						29	31					37	
41		43				47				53					59
61						67			71	73					79
		83						89						97	
101		103				107		109		113					
						127			131					137	139
								149	151					157	
		163				167				173					179
181									191	193				197	199
									211						
		223				227		229		233					239
241									251					257	
		263						269	271					277	
281		283								293					
						307			311	313				317	
									331					337	
						347		349		353					359
361						367				373					379
		383						389						397	

The Sieve of Eratosthanes: all primes less than 400 found.

The Prime Number Theorem

The derivation of the mathematical form of the number of primes less than some number N, usually written $\pi(N)$, was a major problem in Number Theory throughout the 19th century. In the 1790's, Legendre and Gauss independently conjectured that its form is

$$\pi(N) \approx N/\log N$$

where log N is the logarithm of N to the base e (the "natural logarithm"). This conjecture is seen to be reasonable by considering the calculated number of primes up to 1,000,000:

N	$\pi(N)$	$R_N = N/\pi(N)$
10	4	2.5
100	25	4.0
1000	168	6.0...

10000	1229	8.1...
100000	9592	10.4...
1000000	78498	12.7...

The key point to notice from these direct calculations is that in going from one power of 10 to the next, for large N, the ratio R_N increases by approximately 2.3. For example $10.4 - 8.1 = 2.3$ and $12.7 - 10.4 = 2.3$. Now this ratio is remarkably close to the value of $\log_e 10 = 2.30258\ldots$ From this numerical evidence the conjectured value of $\pi(N)$ is

$$\pi(N) = N/\log N \text{ in the limit } N \to \infty.$$

(If $R_N = N/\pi(N) = \log N$ then $R_{N_2} - R_{N_1} = \log N_2 - \log N_1 = \log (N_2/N_1)$ but $N_2 = 10 N_1$ therefore $R_{N_2} - R_{N_1} = \log_e 10 = 2.30258\ldots$).

The calculated growth of $\pi(N)$ with N is compared with the conjectured value in the following table:

N	$\pi(N)$	N/log N	Error (%)
10^3	168	145	13.7
10^4	1,229	1,086	11.6
10^5	9,592	8,686	9.4
10^6	78,498	72,382	7.8
10^7	664,579	620,421	6.6

The predicted value approaches the actual value as N increases.

It took almost 100 years to obtain a rigorous proof of this conjecture, and thereby raise it to the status of a theorem. C. de la Vallee Poussin and Jaques Hadamard independently gave the proof in 1896. Their mathematical analyses involved the most advanced applications of complex number theory available at the time. A more exact value for $\pi(N)$ is obtained by introducing a correction term that uses the celebrated Riemann zeta function. The details of this work are outside the scope of the present talk.

The Random Distribution of Primes

A study of the calculated primes seems to indicate that they are *randomly distributed*. For example in the range 1000 to 1100 we find the primes:

1009, 1013, 1019, 1021, 1031, 1033, 1039, 1049, 1051, 1061, 1063, 1069, 1087, 1091, 1093, and 1097.

There is no obvious pattern to these numbers. This is found to be the case in all ranges so far studied.

The distribution of the spacing between adjacent primes shows interesting variations; the most probable spacing for p less than 10^{100} is 6, this is the first number that is the product of the two primes, 2•3. Relative peaks also occur at 30 (2 • 3 • 5), 210 (2 • 3 • 5 • 7), and 2310 (2 • 3 • 5 • 7 • 11) and so on. For very large primes, the most probable spacing becomes 30.

The following graph shows the result of comparing a theoretical probability distribution, known as a Wigner distribution, with the calculated distribution obtained from 621,940 uniformly distributed random numbers in the interval [0, 1], generated from the primes.

32

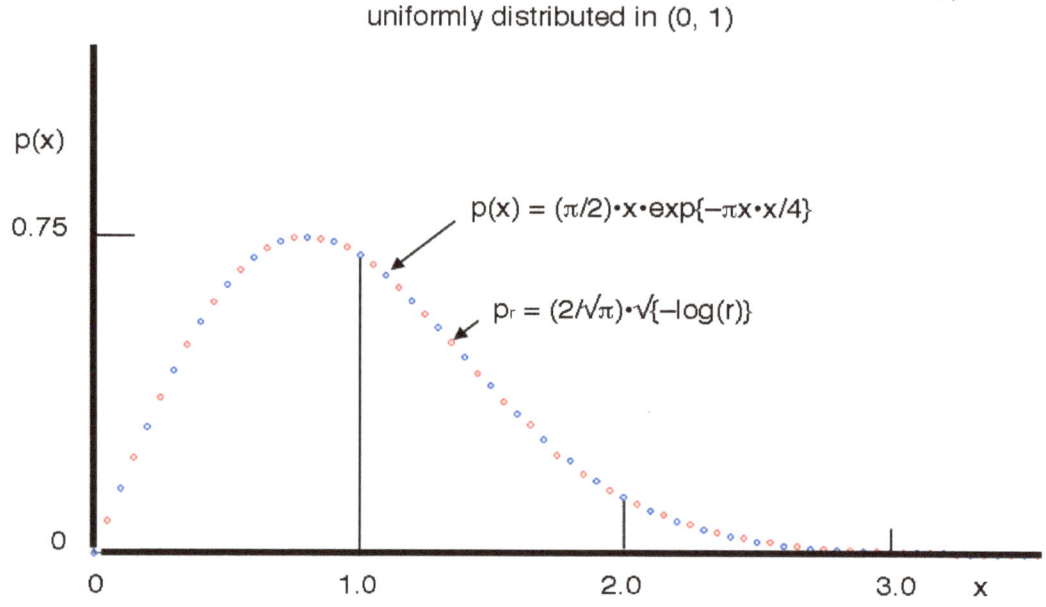

Wigner variates chosen from > 600,000 random numbers, r, uniformly distributed in (0, 1)

$p(x)$

0.75

$p(x) = (\pi/2) \cdot x \cdot \exp\{-\pi x \cdot x/4\}$

$p_r = (2/\sqrt{\pi}) \cdot \sqrt{\{-\log(r)\}}$

0

0   1.0   2.0   3.0   x

The generated curve is in excellent agreement with the theoretical function.

The twin prime conjecture

A challenging question in Number Theory is "are there infinitely many twin prime pairs (p, p + 2), where both members of the pair are prime? "

The first few pairs are (3, 5), (5, 7), 11, 13), (17, 19), (29, 31), (41, 43), (59, 61), ... Numerical evidence supports the conjecture but no rigorous proof has been given. If a proof is found, it will have a profound impact on many outstanding problems in Number Theory, particularly those that involve the famous Riemann hypothesis. Again, the randomness and independence of the primes plays an important part in any discussion of the twin prime conjecture. The following statistical approach shows that the conjecture is reasonable.

The chance of two numbers n and n + 2 both being prime can be considered the same as the chance of getting a head on two successive tosses of a coin. The chance of two successive random and independent events happening is the product of the chance of success of either one; if the probability of one coin coming up heads is ½ then two coins have a probability of ½ • ½ = ¼ of coming up a pair of heads. The prime number theorem states that if n is sufficiently large, and we choose a number n at random in the range o to n, then the chance that n will be prime is about 1/log n. Continuing in this spirit, the chance that both x and x + 2 are prime should be about $1/(\log n)^2$ and therefore there would be about $n/(\log n)^2$ prime pairs between o and n. (This should be corrected to take into account the dependence of x + 2 being prime if x already is prime; this leads to $n/(\log n)^2 \rightarrow (1.32032...) \cdot n/(\log n)^2$). Comparisons between this prediction and numerical calculations gives very good agreement; for example, in the interval 1,000,000,000 and 1,000,150,000 there are 466 prime pairs whereas the formula predicts 461. At this time, the best theoretical result concerning the possible infinitude of twin prime pairs is due to Chen who has proved that *there exist infinitely many pairs p, p + 2 where p is prime and p + 2 is either prime or the product of two primes.*

The calculated spacing distribution between adjacent twin prime pairs for p < 100,000,000 is shown. The variations in the distribution depend on the prime factors of the spacing.

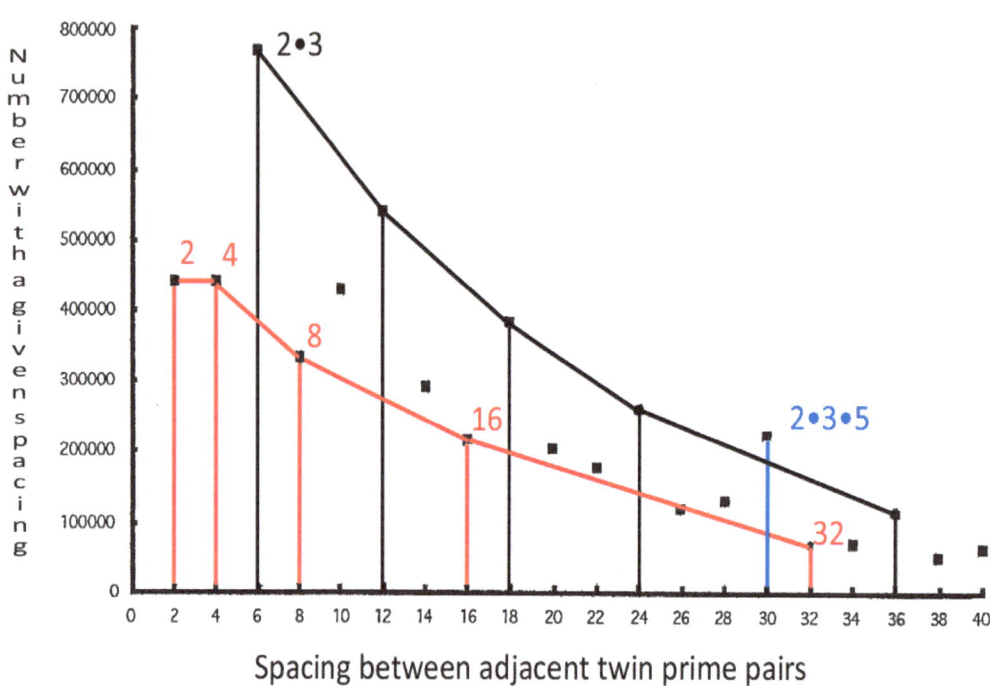

Spacing Distribution of Twin Prime Pairs for Primes p < 100000000

Spacing between adjacent twin prime pairs

The first spacing that is the product of the two different primes 2•3 is almost twice as likely as a spacing of either 2 or 4. For a spacing of 2•3•5 = 30, the distribution increases when compared with the expected value. Increases always occur above the local values whenever they are equal to a "primorial" − 2•3•5•7•11•13•. . .

To observe the distribution for very large spacings requires lengthy calculations. The distribution of spacings for twin primes of the form (11, 13) + 30•m where m = 1,2,3,4 . . . for p < 1,000,000,000 shows the increase at the primorial number 2310.

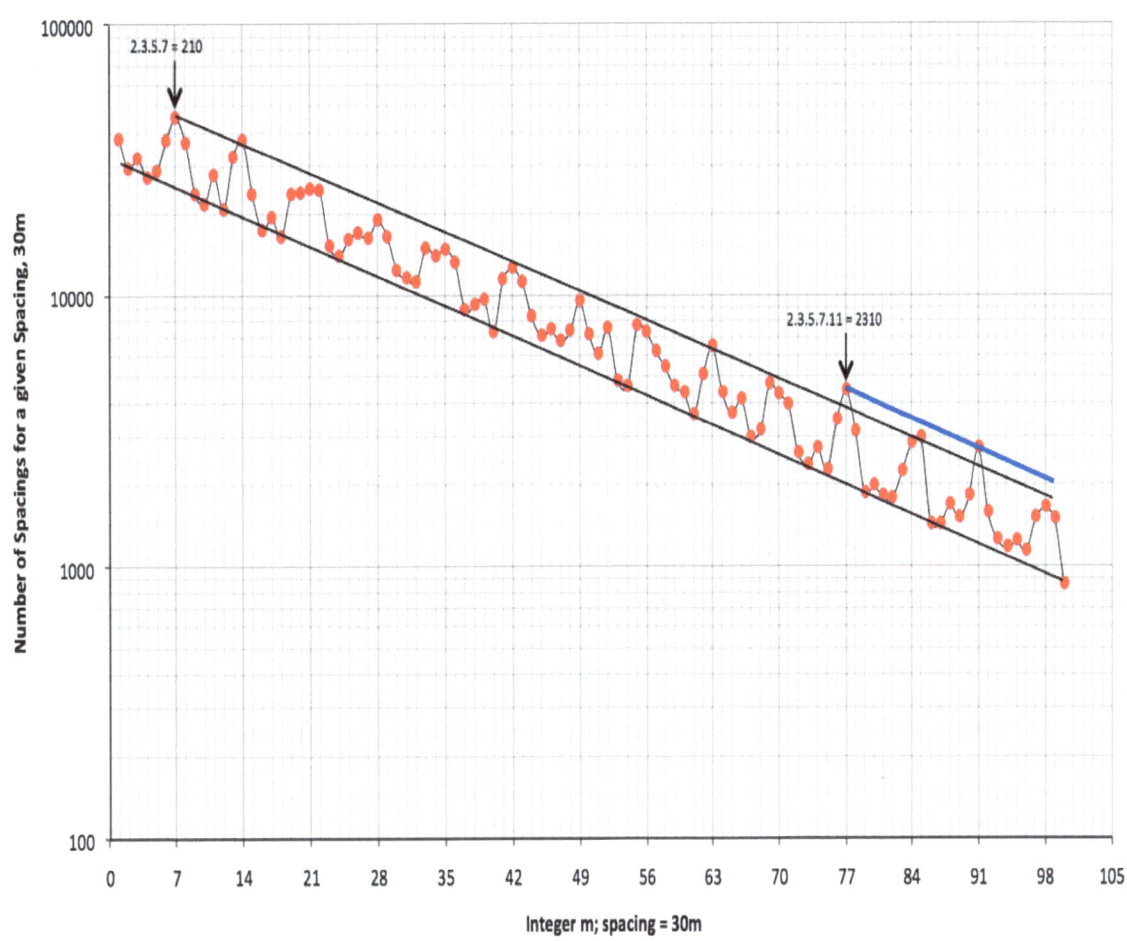

Geometry and Primes

The most important mathematical constant in Geometry is designated π (an abbreviation for the Greek word for perimeter). It is the ratio of the circumference to the diameter, and the ratio of the area to the square of the radius, of all circles. Its value is, to the first few decimal places,

π = 3.145926535 . .

The value of π is currently known to 10^{12} digits.

It is *irrational* meaning that it cannot be expressed exactly as a ratio of integers, m/n, and therefore its decimal representation never ends, and never repeats. It is *transcendental* – no finite sequence of algebraic operations on integers (such as taking roots, summing, taking powers etc.) can equal its value.

In 1737, Euler proved that π can be expressed in terms of repeated products that involve the *primes*:

$$(1/1^2 + 1/2^2 + 1/3^2 + 1/4^2 + \ldots)$$
$$= \Sigma_{n=1,\,\infty}\ 1/n^2$$
$$= 1/(1 - 1/2^2)\cdot 1/(1 - 1/3^2)\cdot 1/(1 - 1/5^2)\cdot \ldots$$
$$= \Pi_{i=1,\,\infty}\ 1/(1 - 1/p_i^2)$$
$$= \pi^2/6$$

This is a most remarkable set of equations in which the *sum* of the inverse squares of *all the integers* is related to a *product* of terms that involve the *primes*, and to the *square of π*.

Euler's proof involves methods that are familiar in High School mathematics; his thought process was, however, far from familiar. He argued as follows:

Consider the infinite (convergent) sum of the inverse squares of the integers

$$z(2) = 1/1^2 + 1/2^2 + 1/3^2 + 1/4^2 + \ldots \tag{A}$$

where $z(2)$ is called the *zeta function* for the index 2.

If we divide throughout by 2^2 we obtain

$$z(2)/2^2 = 1/2^2 + 1/4^2 + 1/\,6^2 + 1/8^2 + 1/10^2 + \ldots \tag{B}$$

Subtracting (B) from (A) gives

$$(1 - 1/2^2)\cdot z(2) = 1 + 1/3^2 + 1/5^2 + 1/7^2 + 1/9^2 + 1/\,11^2 + 1/\,13^2 \,.. \tag{C}$$

Divide (C) throughout by 3^2 to give

$$(1 - 1/2^2)\cdot z(2)/3^2 = 1/3^2 + 1/9^2 + 1/15^2 + 1/21^2 + 1/27^2 + .. \tag{D}$$

Subtracting (D) from (C) gives

$$(1 - 1/3^2)\cdot(1 - 1/2^2)\cdot z(2) = 1 + 1/5^2 + 1/7^2 + 1/11^2 + 1/13^2 + .. \tag{E}$$

We see that in (E), all terms that have 2 or 3 (or both) as factors have been removed. Repeating this process for the infinite number of terms in the sum, gives

. . . . $(1 - 1/11^2)\bullet(1 - 1/7^2)\bullet(1 - 1/5^2)\bullet(1 - 1/3^2)\bullet(1 - 1/2^2)\bullet z(2) = 1$

and therefore

$z(2) = 1/(1 - 1/2^2)\bullet 1/(1 - 1/3^2)\bullet 1/(1 - 1/5^2)\bullet 1/(1 - 1/7^2)\bullet(1 - 1/11^2)$. .

$= \Pi_{i = 1, \infty}\ 1/(1 - 1/p_i^2)$, a product involving all the primes p_i.

Euler gave general versions of the zeta function for different indices that had far-reaching effects on the development of Mathematics and Mathematical Physics. In the mid-19th century, Riemann extended the definition of the zeta function to include indices that are complex numbers.

I have shown that π can be written in terms of an infinite series that includes the *twin primes*, explicitly:

$\pi = 2\sqrt{3}(1 - 2x)$

where

$x = \Sigma_{j=1, \infty}\ (1/[(6j + 1)\bullet(6j - 1)]) = 0.046,550,158$ . . .

(See arXiv:0909.1523v1).

Every twin prime pair greater than (3, 5) is of the form $(6j + 1)\bullet(6j - 1)$.

Some intriguing problems

1.   A remarkable conjecture, known as the Goldbach conjecture (although, in its usual form, it was proposed by the renowned Euler) is:

Every even number is the sum of two prime numbers.

Examples are

14 = 3 + 11, 7 + 7

24 = 5 + 19 , 7 + 17, 11 + 13

This is a particularly challenging problem because it involves *sums and not products of prime numbers*; there is no equivalent fundamental theorem of arithmetic.

2. *Are there infinitely many primes of the form (x^2 + 1)?*

The first few are 5, 17, 37, 101, . . . No one knows how to write down the first line of a proof!

3. *Searching for patterns*

Following the early work of Ulam, Robert Sacks created a *number spiral* in which all the positive integers are placed, equidistantly, along the arms of a spiral. The patterns associated with prime numbers show interesting variations in the density of primes along each arm. The arms are characterized by the quadratic forms $x^2 + x \pm N$ where $N = 0, 1, 2, 3, \ldots$

The Robert Sacks' Number Spiral

The prime numbers are shown in purple. The variations in the density of primes along each arm are not understood. The squares of numbers, associated with the single arm x^2, are shown in red.

The Diffie-Hellman Method (1976)

(First invented by Williamson, a member of the British Secret Service, in 1974. His work was not declassified until 1999)

For a positive integer n, two integers a and b, are said to be congruent modulo n, if a • b is an integer multiple of n. The number n is the modulus of the congruence. For example, 38 is congruent to 14 mod 12 because 38 - 14 = 24 = 2•12. *Modular arithmetic is related to the remainder in division. Finding the integer remainder is referred to as the modulus operation.* We write, for example $10 = 2^5$ mod 11 because 2^5 /11 = 32/11 = 2 and 10/11 remainder; the integer remainder is therefore 10.

In practice, the exchange of a secret number is carried out as follows:

Let two distant persons **A** and **B** agree to use a common prime number p = 11 and "base" g = 2. These are exchanged via a *public network*.

A chooses a secret integer a = 5, and sends $N_a = g^a$ mod p = 2^5 mod 11 = 10 to **B**.

B chooses a secret integer b = 7, and sends $N_b = g^b$ mod p = 2^7 mod 11 = 7 to **A**.

A computes 7^5 mod 11 = 10, and **B** computes 10^7 mod 11 = 10.

Both **A** and **B** have obtained the same secret number, 10. If a, b and p are large numbers, more than 100 digits in each, it is impossible, with all the computing power available, to find a or b given g, p, and N_a and N_b.

NEW DIMENSIONS: A TRIBUTE TO BENOIT MANDELBROT (1924 – 2010)

November 14, 2010

To honor the achievements of our late, distinguished colleague, Benoit Mandelbrot, Sterling Professor Emeritus of Mathematical Sciences at Yale, I offer this overview of the field for which he will be remembered, namely *Fractal Geometry*. It is no exaggeration to say that no field of Mathematics has so captured the imagination of the public at large. The ready availability of sophisticated computer programs has made it possible for all of us to become expert graphics designers. Before describing the key aspects of *fractals* – a word coined by Mandelbrot – I shall show some typical images to illustrate their varied and often stunning shapes.

Many examples appear in Mandelbrot's pioneering book *The Fractal Geometry of Nature*. His work was wide-ranging, covering irregular and often chaotic natural phenomena in the physical, biological, and social sciences. It was not by accident that his major discoveries were made during his long tenure as an IBM Fellow. Computed fractals are the logos of the Computer Age in which we now live. I shall place his work in its historical context, and I shall discuss the key mathematics involved at a level that is not too challenging.

We begin by discussing the meaning of "dimension". Experience leads to the notions that a line has a dimension of 1 – only one number is needed to define a point on it. An area, such as a tabletop has a dimension of 2 – two numbers are required to specify a point in a plane, and a cube or sphere has a dimension of 3 – three numbers are required to specify a point in a cubical or spherical volume. As an example, we may use the 3 Cartesian coordinates x, y, z to define a point in 3-space.

We now take a different approach in defining "dimension"; it involves the ideas of "*magnification*" and "*self-similarity*". We shall consider the magnification by 2 of a line, a square, and a cube:

A line of length L becomes a line of length 2L; it consists of two self-similar lines each of length L.

A square of area A becomes a square of area 4A; it consists of four self-similar areas each of area A.

A cube of volume V becomes a cube of volume 8V; it consists of eight self-similar cubes each of volume V.

We summarize these results in the following table

Figure	Magnification m	Number of self-similar objects, N	Dimension D
Line	2	2	1
Square	2	4	2
Cube	2	8	3

We see that in each case we can write the equation $N = m^D$.

Taking logarithms, we obtain

$$\log N = D \cdot \log m$$

so that the dimension is given by

$$D = \log N / \log m;$$

(In the case of the cube, $N = 8 = 2^3$ therefore $D = \log 8 / \log 2 = 3$, as it should)

This approach, that seems to complicate our traditional ideas concerning dimensions, nonetheless leads us to ask questions concerning possible magnifications and self-similarities that result in a value of D that is non-integer. Two important cases were discovered about a century ago; they are known as Cantor "dust" and the Sierpinski triangle. They play a major role in the development of fractals.

Cantor's dust:

In the late 1800's, Cantor introduced ideas that challenged the very foundations of classical geometry. His argument went as follows.

Consider a line of length L:

$$L$$
_____

Remove the middle third to generate two lines, each of length L/3.

L/3   L/3
_____   _____

We now have two self-similar lines and each one would need to be magnified by x3 to obtain the original line.

Cantor continued this process by removing the middle third of each line segment:

L/9   L/9   L/9   L/9
___   ___   ___   ___

and so on, an infinite number of times.

The final result is a set of infinitely small line segments that are not perfect points; they are self-similar versions of the original line of definite length L. In this example, we have, at each step (iteration) a magnification, m = 1/3 and a number of self-similar objects, N = 2.

The dimension associated with the Cantor set is therefore

D = log N/log m = $-$ log 2/ log (1/3) (the minus sign is introduced because m < 1)

= 0.69314732.../1.0986123... = 0.6309299...

The dimension is somewhere between that of a point (D = 0) and that of a line (D = 1).

The Sierpinski triangle

In the early 1900's, the Polish mathematician Sierpinski constructed a recurrence based on equilateral triangles.

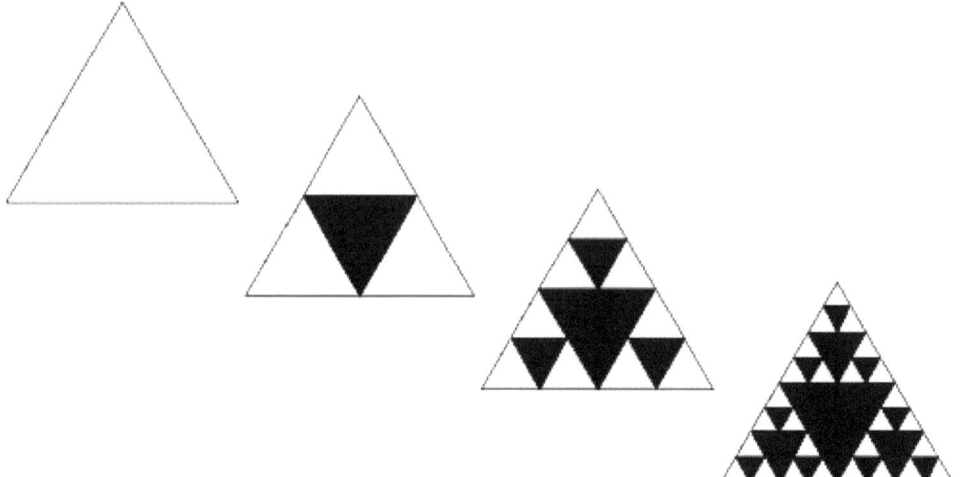

At each iteration, the equilateral triangle formed from the midpoints of the sides is removed. This process is repeated infinitely many times. The number of self-similar triangles created by the first iteration is 3, and the magnification is ½ (each side is halved in length). The dimension is therefore

$$D = - \log 3 / \log (1/2)$$
$$= 1.0986123... / 0.6931479...$$
$$= 1.5849625...$$

The Sierpinski iteration results in a paradox: the area being subtracted from a particular area becomes smaller and smaller, and therefore the remaining area approaches a certain fraction of the original area. Endless iterations of the removal of middle triangles results in an increase in the length needed to enclose the divided area without bound – infinite length encloses finite area – not a result consistent with Euclidean geometry. The formal development of geometry that involves non-integer dimensions was due to Hausdorff (1919) and Besicovich (1935). The "dimension" that involves the ratio log N/log m is known as the Hausdorff – Besicovich dimension.

The process of numerical iteration has been important in Mathematics for many centuries. The following example illustrates Newton's method for finding the square root of a number n.

Newton's iteration

To find the square root of a number n the iterative process

$$x_{k+1} = \tfrac{1}{2}\{x_k + n/x_k\}$$

is carried out over-and-over again until the result converges on the square root of n. The starting value of x_k is taken to be $x_0 = 1$. The following table lists the successive iterated values of $\sqrt{50}$.

k + 1	"$\sqrt{n}$"
1	25.5
2	13.730392
3	8.685974
4	7.22119
5	7.072628
6	7.071068
7	7.071068

In this particular case, the value obtained for the square root of 50 is remarkably accurate after 7 iterations.

Parabolic iterations

An example of the iterative process that will be useful in later discussions involves the real quadratic form

$$y = f(x; d) = x^2 + d$$

where d is a real parameter.

We label each iteration c_n, n = 1, 2, . . . If the initial value of $x = x_0$ then

$$c_1 = x_0^2 + d$$
$$c_2 = c_1^2 + d$$
$$c_3 = c_2^2 + d$$
$$. . .$$

46

As an example, if $d = -0.5$, and $x_0 = 1$ we obtain the following iterative values:

$c_1 = 0.5$, $c_2 = -0.25$, $c_3 = -0.4375$, $c_4 = -0.3086$, $c_5 = -0.4048$, $c_6 = -0.3362$
The sum $(c_5 + c_6)/2 = -0.370$. After many iterations, it is found that c_n converges on the number $x = -0.366$. This value corresponds to one solution of the quadratic equation $x^2 - x - 0.5 = 0$. The other solution is 1.366. For all initial values of x in the range $[-1.366, 1.366]$, and for $d = -0.5$, the iterative process results in the value $c_\infty = -0.366$; the set of iterated values form a "filled-in" set (see following figure). If the selected initial values x_0 are not in the range $[-1.366, 1.366]$ then the associated iterations lead to values of c_∞ that tend to infinity. (The range is found by finding the intersections of the quadratic $x^2 - 0.5$ and the line $y = x$).

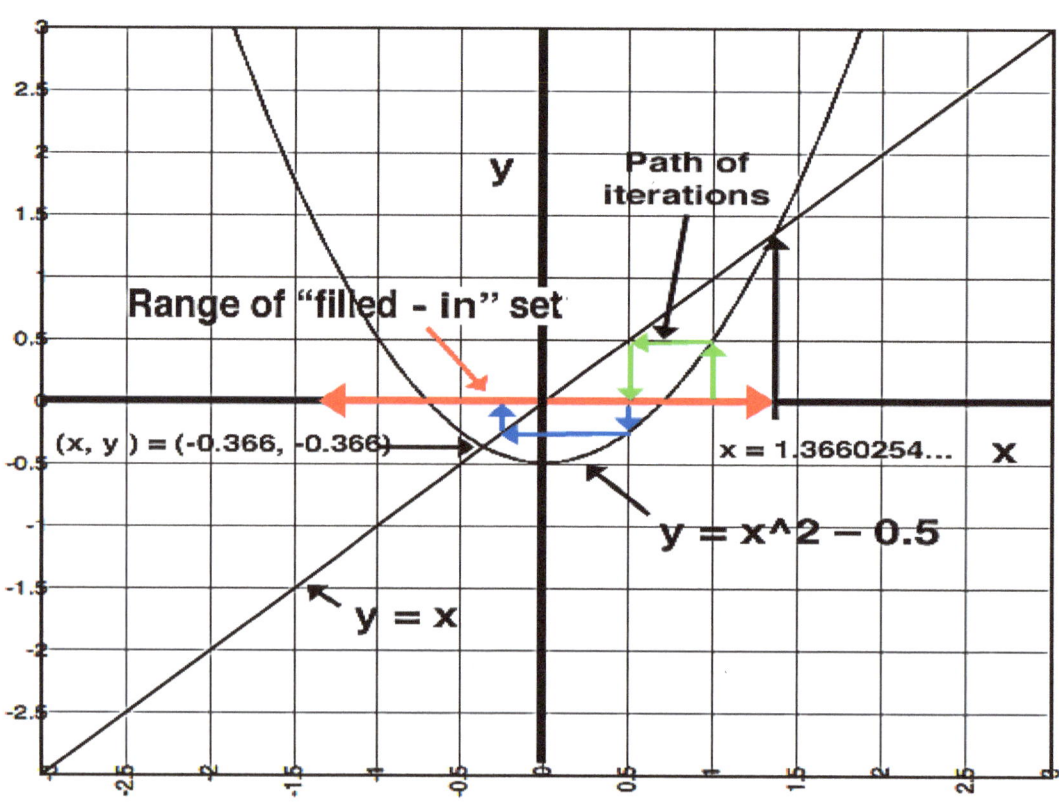

For different values of the parameter d, new sets of iterates are obtained, and new valid ranges apply.

In studying iterations, it may be necessary to perform many calculations. This was not possible until the development of powerful, high-speed computers.

New iterative functions

During his education in Paris, Mandelbrot was influenced by two eminent mathematicians, Paul Levy and Gaston Julia; their research interests had a lasting effect on his work.

Fatou (1917)/Julia (1918) sets

Fatou and Julia were concerned with iterations of quadratic forms of the type

$$z \rightarrow z^2 + c$$

in which both z and the parameter c are, in general, *complex numbers*.

The following brief account of the properties of complex numbers will be useful in developing the mathematical basis of fractals. The account will begin with an introduction to a topic in linear algebra:

Linear transformations and matrices

Let a point P[x, y] in a Cartesian frame be *rotated* about the origin through an angle of 90°; let the new position be labeled P'[x', y'].

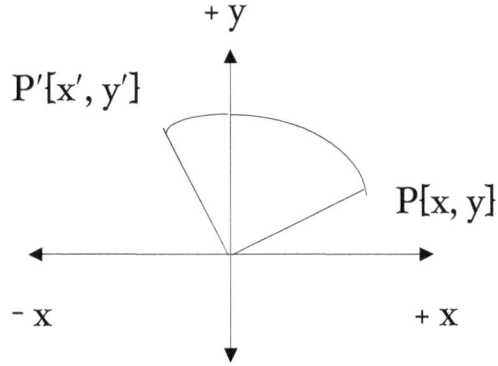

48

We see that the new coordinates are related to the old coordinates as follows:

x' (new) = $-y$ (old) = $0 \cdot x - 1 \cdot y$

and

y' (new) = $+x$ (old) = $1 \cdot x + 0 \cdot y$

where we have written the x's and y's in different columns for reasons that will become clear, later.

Consider a *stretching* of the material of the plane such that all x-values are doubled and all y-values are tripled. The old coordinates are related to the new coordinates by the equations

y' = $3y = 0 \cdot x + 3 \cdot y$

and

x' = $2x$ = $2 \cdot x + 0 \cdot y$

These are particular examples of the more general transformation

$x' = a \cdot x + b \cdot y$

and

$y' = c \cdot x + d \cdot y$

where a, b, c, and d are real numbers.

In the above examples, we see that *each transformation is characterized by the values of the coefficients, a, b, c, and d:*

For the *rotation* through 90°:

a = 0, b = −1, c = 1, and d = 0;

and for the 2x3 *stretch*:

a = 2, b = 0, c = 0, and d = 3;

In the 1840's, Cayley recognized the *key role of the coefficients in characterizing the transformation* of a coordinate pair [x, y] into the pair [x′, y′]. He therefore "separated them out", writing the *pair of equations in column-form*, thus:

$$\begin{vmatrix} x' \\ y' \end{vmatrix} = \begin{vmatrix} a & b \\ c & d \end{vmatrix} \begin{vmatrix} x \\ y \end{vmatrix}$$

This is a *single* equation that represents the original *two* equations. We can write it in the symbolic form:

P′ = MP ,

which means that the point **P** with coordinates x, y (written as a *column*) is changed into the point **P′** with coordinates x′, y′ by the operation of the **2 x 2 matrix operator M**.

The *matrix* **M** is

$$\mathbf{M} = \begin{vmatrix} a & b \\ c & d \end{vmatrix}$$

The algebraic rule for carrying out the "matrix multiplication" is obtained directly by noting that the single symbolic equation is the equivalent of the two original equations. We must therefore have

$$x' = a \text{ times } x + b \text{ times } y$$

and

$$y' = c \text{ times } x + d \text{ times } y.$$

We *multiply rows of the matrix* by *columns of the coordinates*, in the correct order.

The matrix operator that rotates P through 90° is

$$R(90°) = \begin{vmatrix} 0 & -1 \\ 1 & 0 \end{vmatrix}$$

If we rotate P' through another 90° the point P" with coordinates [−1, 0] is obtained. The double operation means that we must calculate $R(90°) \cdot R(90°)$. Multiplying rows by columns of the matrices we find

$$R^2(90°) = \begin{vmatrix} -1 & 0 \\ 0 & -1 \end{vmatrix} = -\mathbf{I} \text{ where } \mathbf{I} \text{ is the unit matrix.}$$

The unit matrix leaves all points unmoved; it is the equivalent of the number 1. Taking the square roots, we obtain

$$R(90°) = \sqrt{-1} = i$$

A rotation through 90° is equivalent to "operating" with **i**.

In general, a complex number z has two parts, a real part x and an imaginary part y, which we write

$$z = x + iy$$

The rules for adding and multiplying complex numbers are straightforward. It is only necessary to use the fact that $i^2 = -1$, etc. The rules are illustrated in "Argand" diagrams (the complex plane):

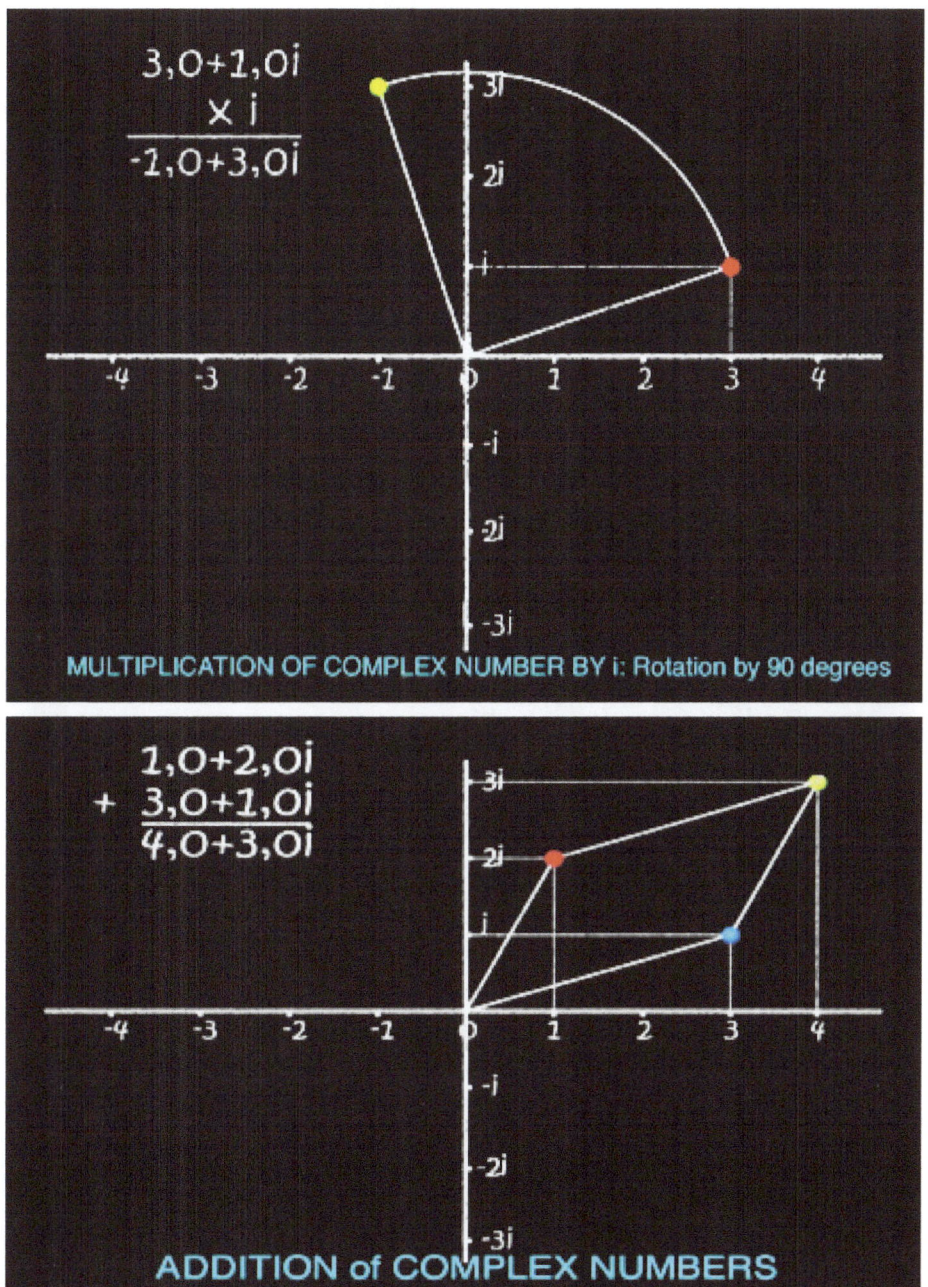

MULTIPLICATION OF COMPLEX NUMBER BY i: Rotation by 90 degrees

ADDITION of COMPLEX NUMBERS

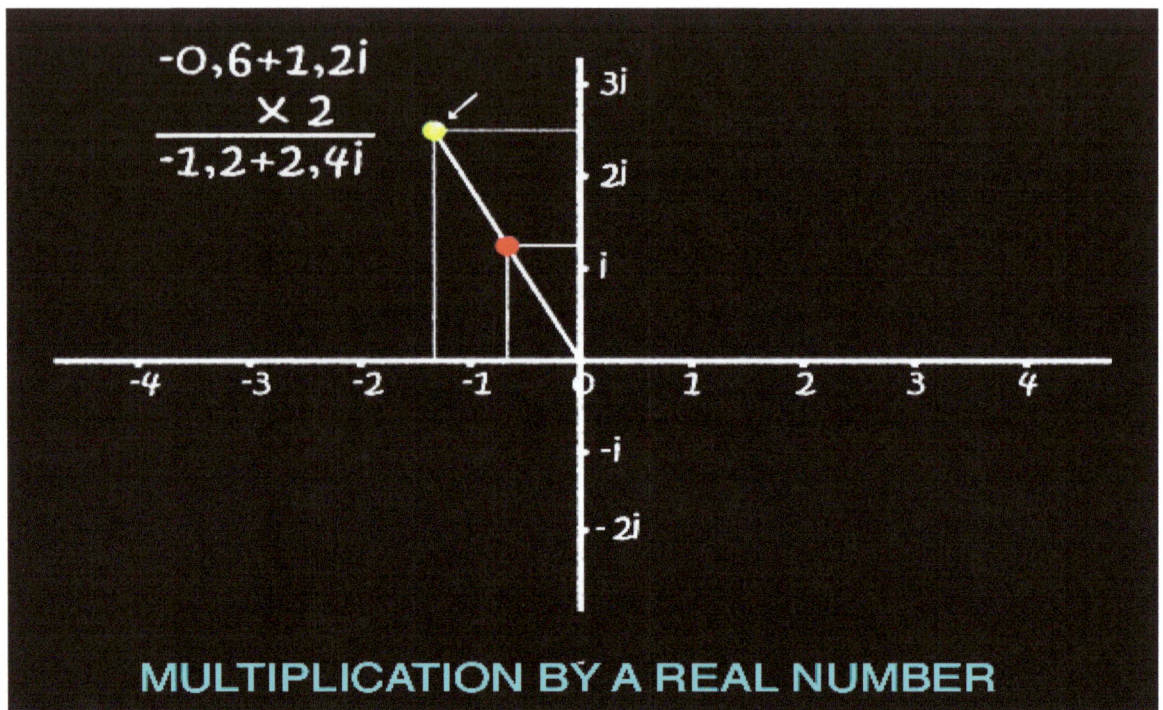

MULTIPLICATION BY A REAL NUMBER

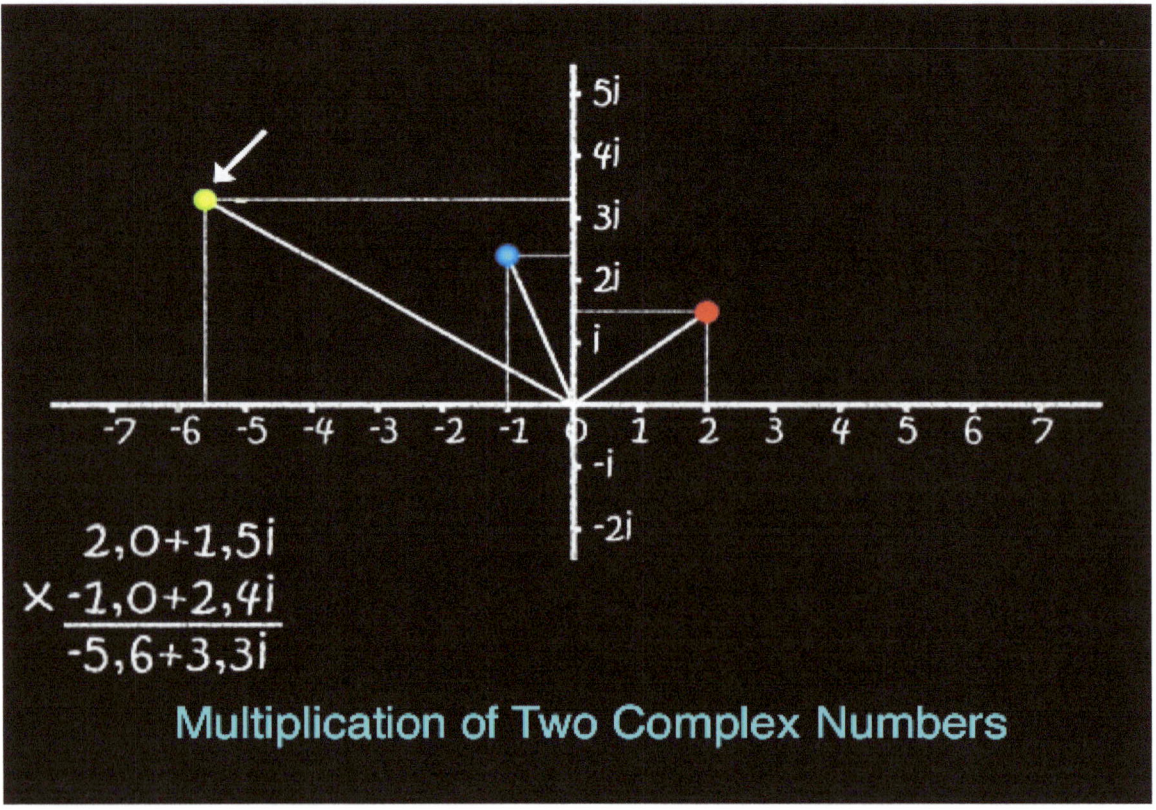

Multiplication of Two Complex Numbers

The length of the line from the origin to a point in the complex plane is called the *modulus*, and the angle made by the line with the x-axis is called the *argument* of the complex number. These quantities will be shown to be useful in the calculation of fractals.

We can now calculate the first few (iterated) terms in a typical Julia set. Let $z = 2 + i3$ and $c = 1 + i1$ then $z^2 + c$ is

$$(-5 + i12) + (1 + i1) = -4 + i13$$

Iterating, we obtain

$$(-4 + i13)^2 + (1 + i1) = -152 - i103,$$

iterating again gives

$$(-152 - i103)^2 + (1 + i1)$$

This process is repeated many times. In this way, an "*orbit* " of the iterated values of z is obtained for a given value of the parameter c. Different orbits are obtained for different values of c. In practice, it is found that certain orbits go to infinity, while others go to well-defined values. (Recall the earlier example of the real quadratic function and the limited range of the "filled-in sets"). In order to limit the number of iterations made, the calculation is stopped if the modulus of the complex number, after many iterations, exceeds a given value. The number of iterations needed to reach the chosen maximum modulus is calculated and the associated point c is colored accordingly. An image of colored points is thereby built up. I have calculated the orbits of points in Julia sets for values of the numbers x, y, u, and v in $z = (x + iy)^2 + (u + iv)$:

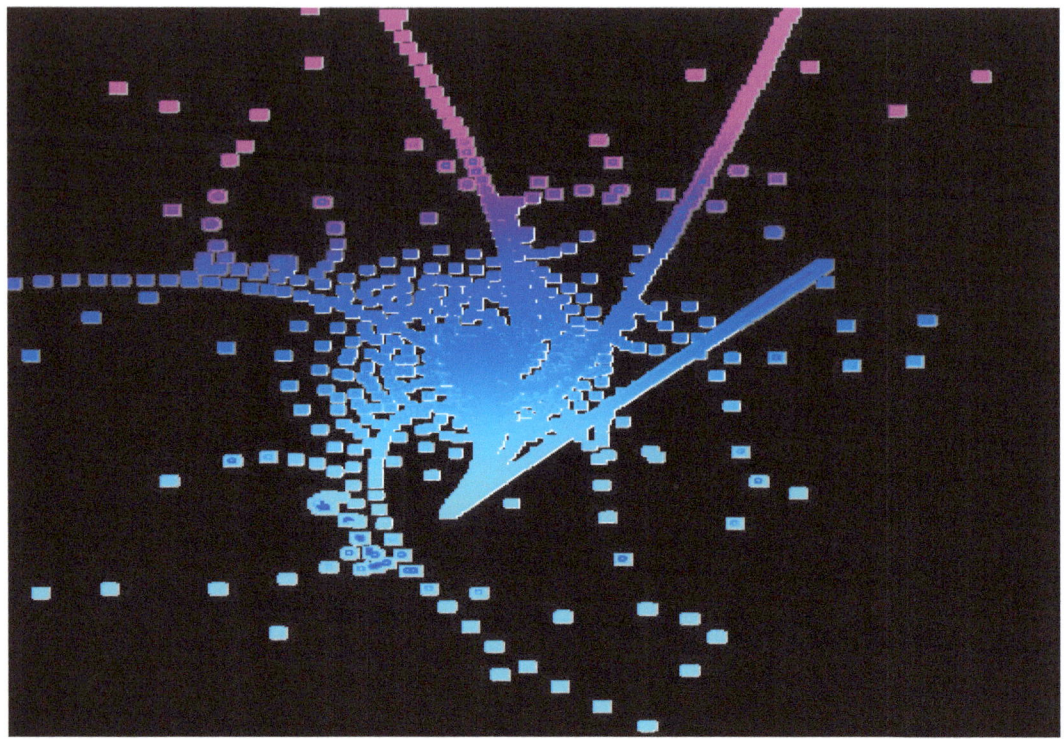

The image is colored for display purposes. Those orbits that go off to infinity are clearly seen. The Julia set has (real) coordinates $(x^2 - y^2 + u, 2xy + v)$.

The Mandelbrot set

The Mandelbrot set is related to the Julia set; it iterates the complex function $c \rightarrow c^2 + c$ ($z_0 = 0$). Points for which the orbit does not tend to infinity are in the set. The set is characterized by the following shape in the complex plane in which points are colored according to the number of iterations required to reach the modulus, $r_{max} = 2$. At all levels of magnification the shapes are closely self-similar.

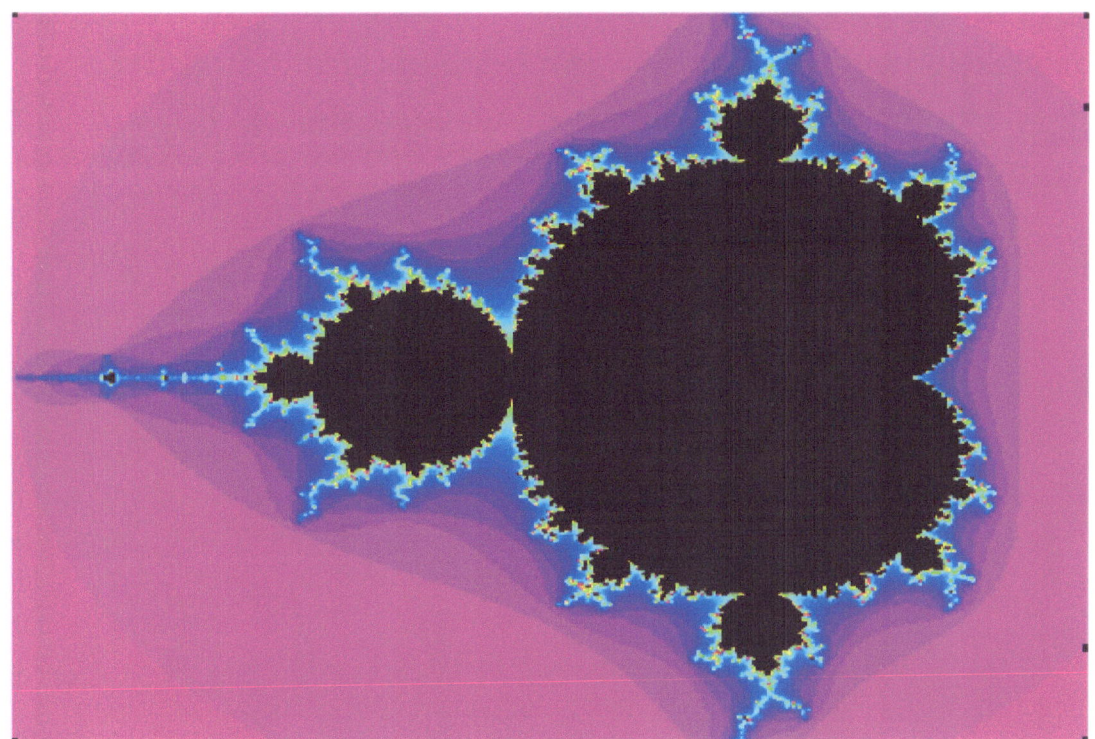

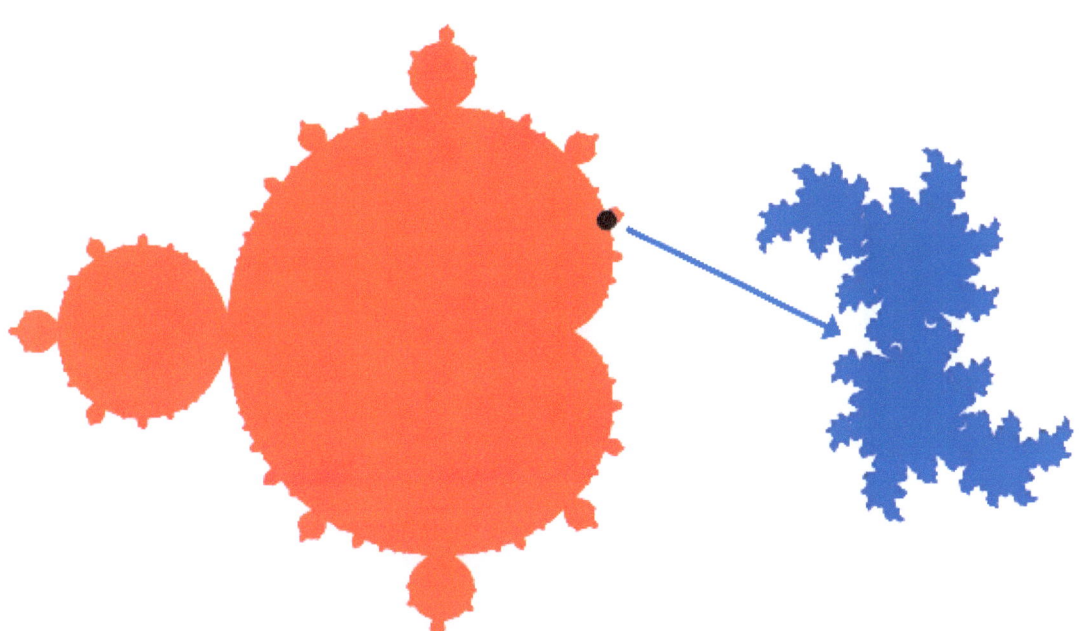

Julia set associated with the point on the Mandelbrot set

For certain values of c on the surface of the Mandelbrot set, the Julia set explodes, and all points go to infinity – no Julia figure remains!

Exact self-similarity is found only in those cases that involve Euclidean geometric forms. The "fractal" forms exhibit near-self-similarity. Examples of the ubiquitous nature of fractals are shown.

Fractal fern showing increasing number of iterations.

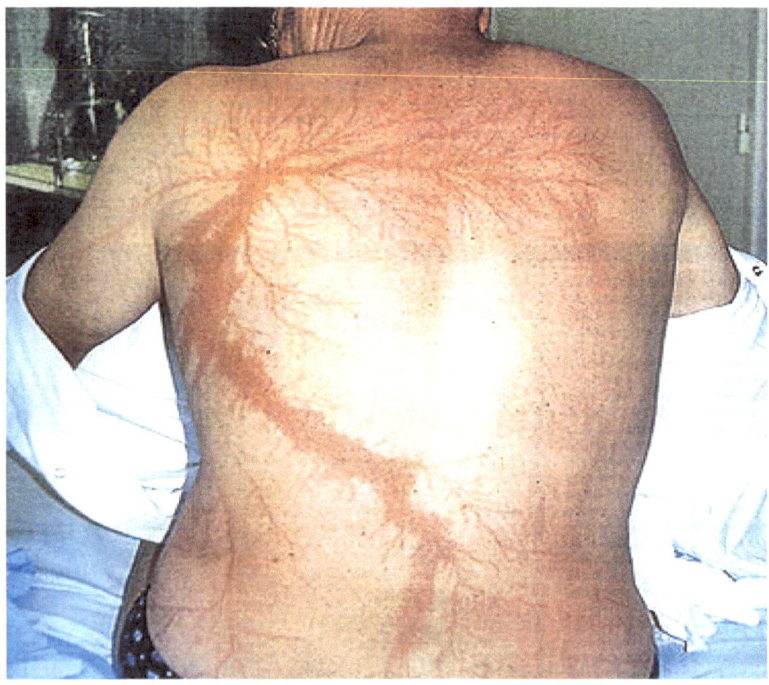

Lightening strike – a natural fractal

Hydrothermal spring – a natural fractal

Natural Tibet mountain fractals

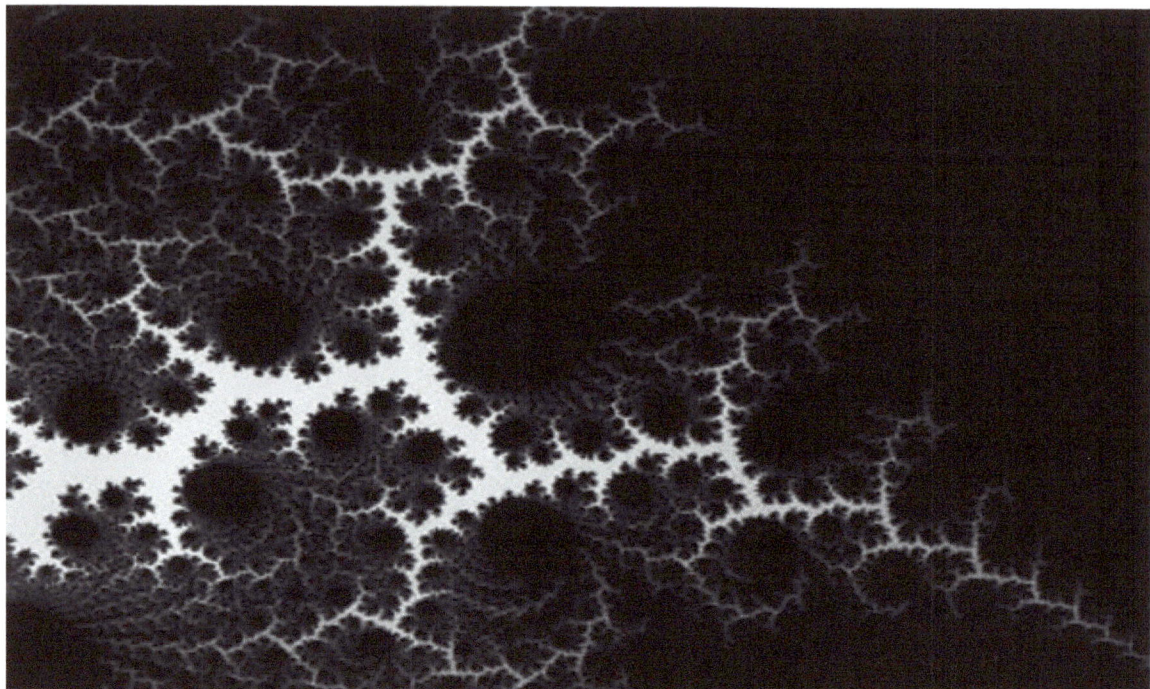

Mathematical mountain fractal

A mathematical landscape: fractal images are widely used in the Film Industry.

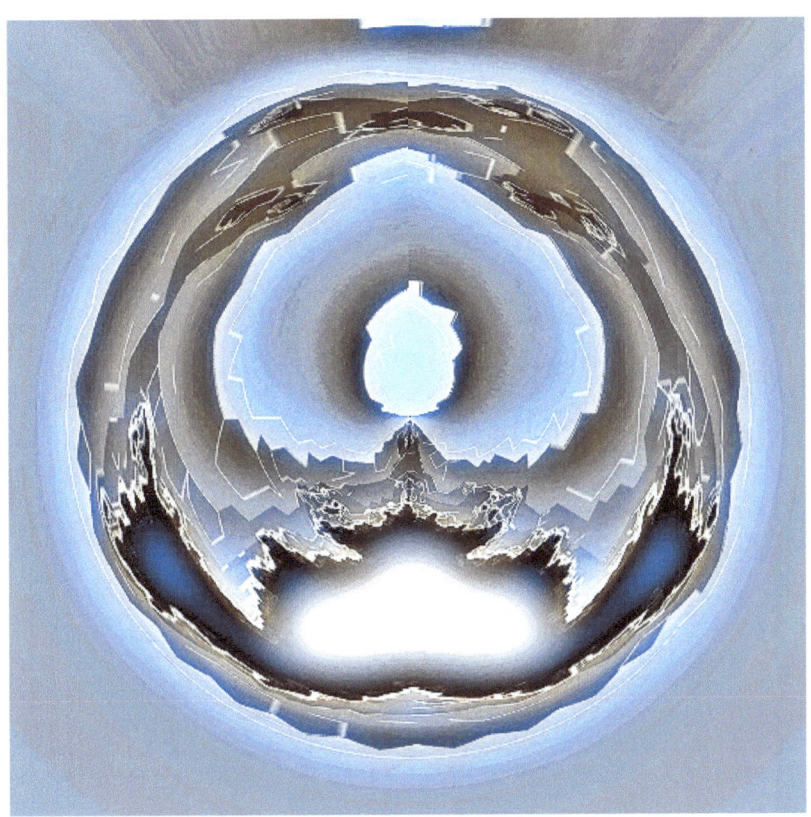

I created this image by calculating the orbits associated with a complex quartic equation, followed by a coordinate transformation..

Mandelbrot's major contribution was to recognize, exploit, and champion the *universality* of fractal geometry. He was the first to apply the ideas to the analysis of fluctuations in stock market prices, to the analyses of dynamical systems, and to the study of chaotic phenomena. In 1970, he gave the Trumbull Lecture at Yale entitled *Statistical Self Similarity and Very Erratic Chance Fluctuations*. He emphasized the inadequacy of traditional geometry in dealing with complex structures that are not continuous and do not have classically-defined shapes. His work continues to have a major influence on the formal development of Topology. Here at Yale, we were fortunate to have him as a colleague during the last two decades of his extraordinary life.

Addendum

The following example illustrates an iterative process that is related to the Julia set.

Iteration of Real Quadratics and the Corresponding Complex Quadratic.

Consider the iteration of the real quadratic forms

$$x_n \to a x_{n-1}^2 + 2g x_{n-1} y_{n-1} + b y_{n-1}^2 + u$$
$$y_n \to c x_{n-1}^2 + 2h x_{n-1} y_{n-1} + d y_{n-1}^2 + v$$

where a, b, c, d, g and h are real coefficients, and u and v are real parameters that are constant for a definite iterative sequence.

If $a = 1$, $b = -1$, $c = 0$, $d = 0$, $g = 0$, and $h = 1$ then

$$x_n \to x_{n-1}^2 - y_{n-1}^2 + u$$
$$y_n \to 2 x_{n-1} y_{n-1} + v.$$

These particular forms are interesting because they are identical to those obtained by studying the complex iteration

$$z_n \to z_{n-1}^2 + c$$

where (dropping the subscripts)

$$z = x + iy \text{ and } c = u + iv.$$

(We have

$$(x + iy)^2 = x^2 - y^2 + 2ixy$$

and therefore

$$z = (x^2 - y^2 + u) + i(2xy + v).$$

In the complex plane, the coordinates of z are $[(x^2 - y^2 + u), (2xy + v)]$.

If, for a given c, the recurrence

$$z_n \to z_{n-1}^2 + c$$

does not tend to infinity, the starting point, z_0 belongs to the "filled-in" Julia set.

PERSPECTIVE AND BEYOND

This lecture is presented as a PowerPoint slide show.

Abstract

Perspective transforms our world of real, 3-dimensional objects into images in 2-dimensions. In this lecture, the development of *perspective* in drawing and painting, from the works of early Chinese artists to the formal approach of artists during the Renaissance, is discussed.

Moving beyond traditional perspective, the subjects of *projective geometry* (Desargues, 1610), *matrix transformations* (Arthur Cayley,1840), and *conformal transformations* (Joukowski, 1906) are introduced at a mathematical level typical of that achieved in high school. An example of the practical use of abstract (conformal) transformations is given that reconnects the subject to its everyday roots.

Zhang Zaduan
Along the River During the Quingming Festival
Beyond the City walls: a section of the panorama

Zhang Zaduan
A section of the panorama *"Along the River . . "*
The City gate

Zhang Zaduan, painted about 1120.
A section of *"Along the River During the Quingming Festival "* showing the Rainbow Bridge, boats, buildings, and activities of the Song people.

The panorama is 17 feet long and 1 foot high

Ambrogio Lorenzetti (1332)
"St Nicholas giving dowry to poor girls"

A. Lorenzetti (1340)
"Effects of Good Government . ."

Pierro della Francesca (1465)
The Duke of Urbino
The artificial landscape is in *linear, color* and *detail* perspective.

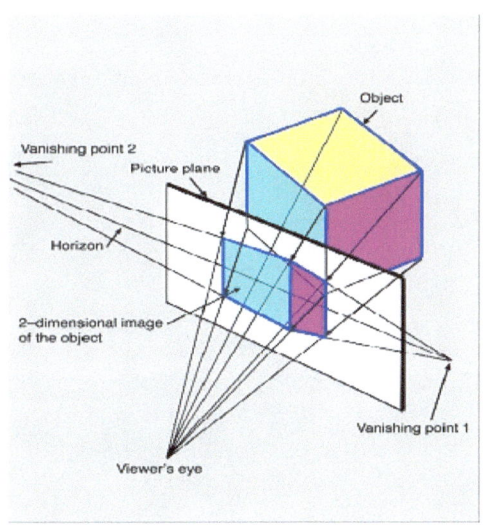

Perspective image of a cube obtained by following rays of light from the object to the
viewer's eye. Intersections of the rays with the picture plane are painted.

Newbury Bridge, Berkshire   F. W. K. F. 1964/2005

Lines of perspective. The lines of the lock-gate lever, its iron support,
and the tow-path disturb the primary perspective of the viewer.

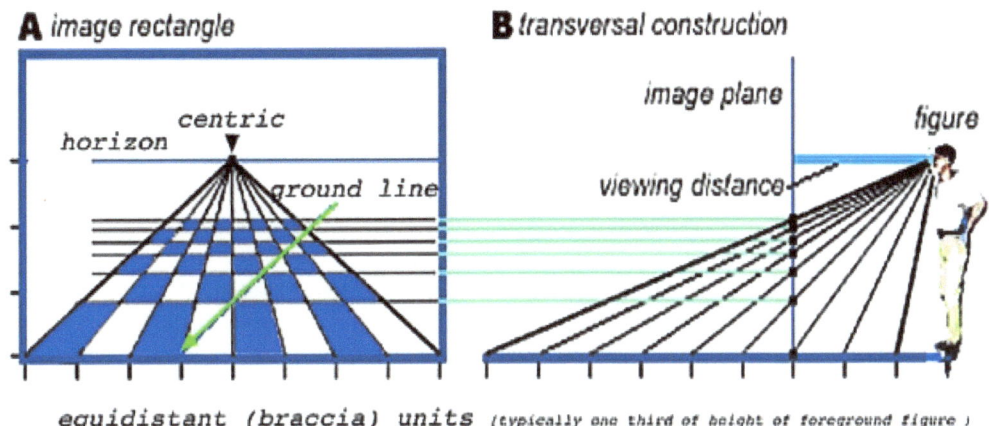

Alberti's construction (1435)

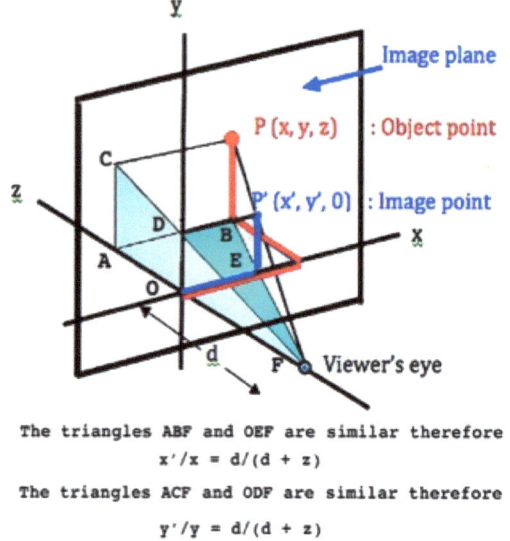

The triangles ABF and OEF are similar therefore

$$x'/x = d/(d + z)$$

The triangles ACF and ODF are similar therefore

$$y'/y = d/(d + z)$$

These are the **mapping equations** x -> x' and y -> y'.

Geometric analysis of the projection of a point P onto the image plane, giving the point P'

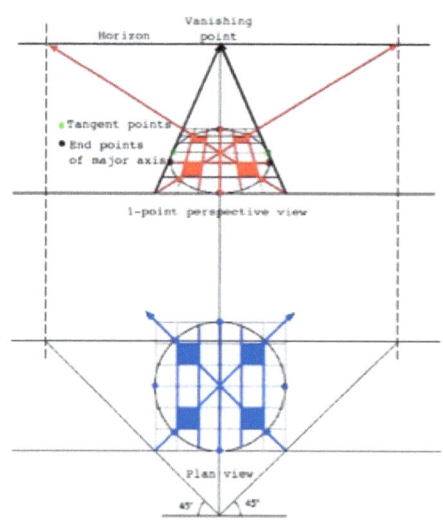

Perspective view of the circle is an ellipse

Alberti's demonstration of the perspective view of a circle

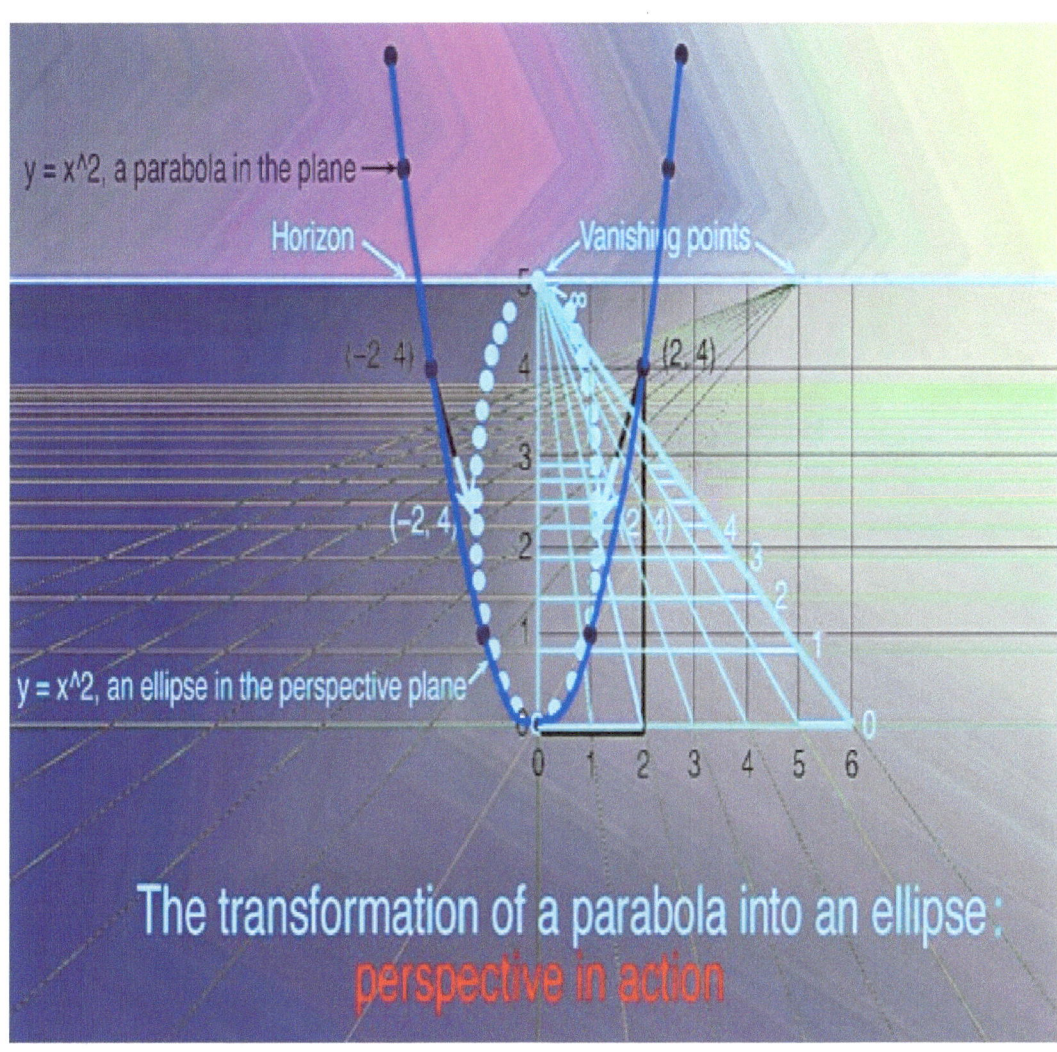

The transformation of a parabola into an ellipse: perspective in action

Perspective view of a cube: a new geometry

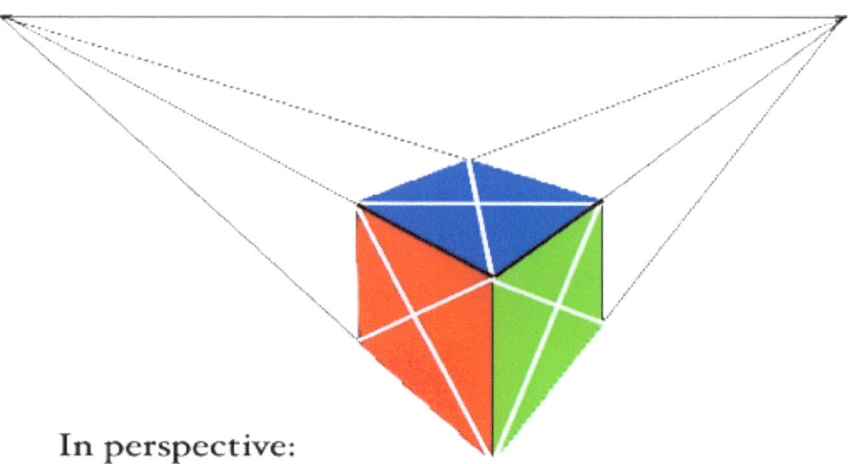

In perspective:
1. lengths change; 2. angles change; 3. parallel lines converge and 4. ratios of lengths not preserved

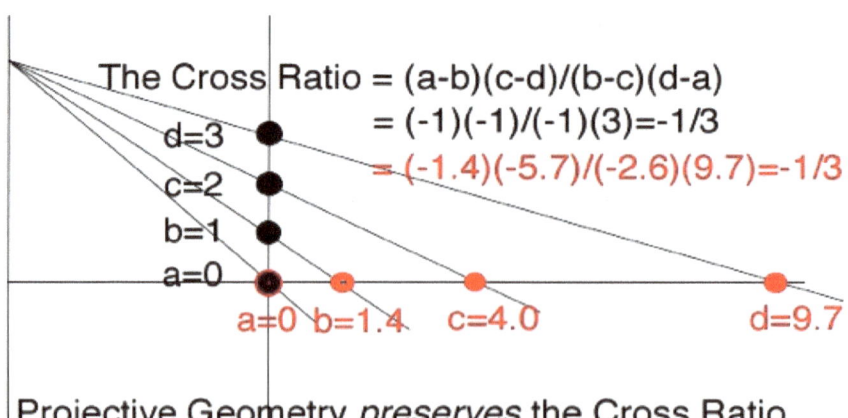

The Cross Ratio $= (a-b)(c-d)/(b-c)(d-a)$
$= (-1)(-1)/(-1)(3) = -1/3$
$= (-1.4)(-5.7)/(-2.6)(9.7) = -1/3$

Projective Geometry *preserves* the Cross Ratio
(Pappus' Theorem, c. 300 A.D.)

The formal theory of perspective was developed by **Desargues** in the early 1600's. *Euclidean Geometry* involves not only points and lines but also lengths and angles. Desargues' *Projective Geometry* does *not* involve lengths and angles. He discovered the following theorem:

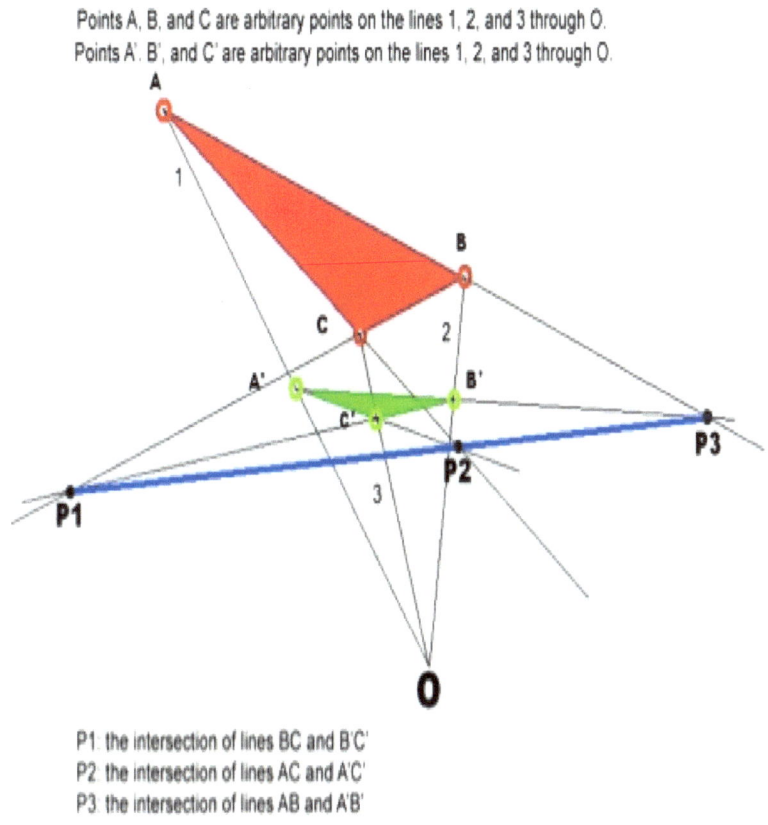

Points A, B, and C are arbitrary points on the lines 1, 2, and 3 through O.
Points A', B', and C' are arbitrary points on the lines 1, 2, and 3 through O.

P1: the intersection of lines BC and B'C'
P2: the intersection of lines AC and A'C'
P3: the intersection of lines AB and A'B'

Desargues' Theorem: the points P1, P2, and P3 are collinear.

The two triangles are "point perspective", if and only if, they are also "line perspective".

Proof of Desargues' Theorem

The proof is straightforward if the problem is considered in 3-dimensions, as shown:

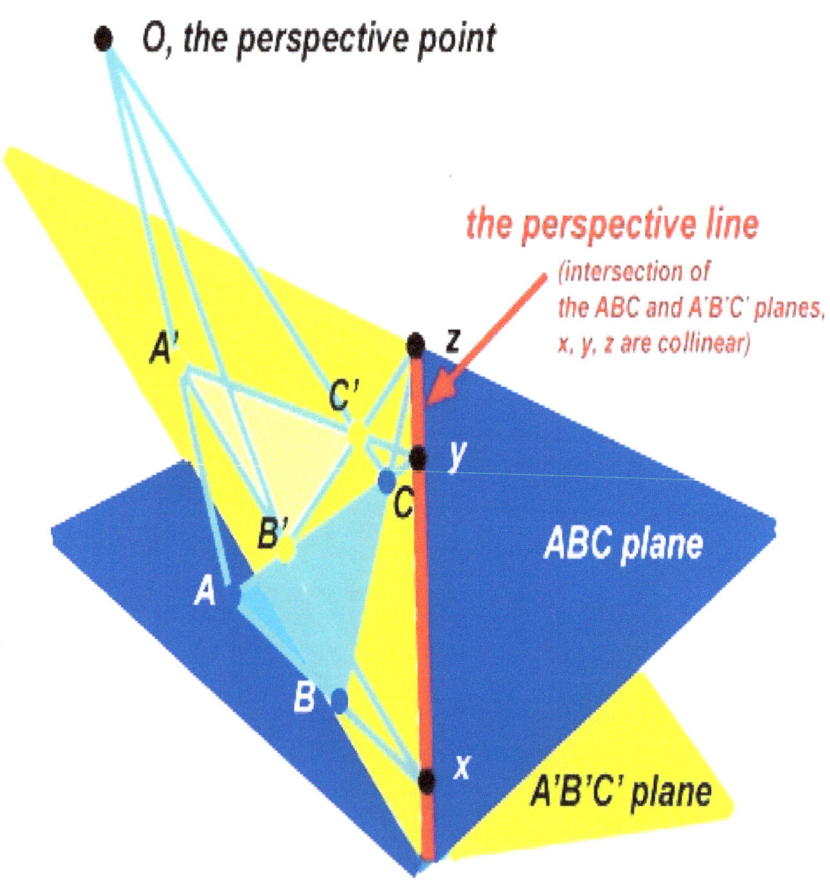

Let a point P[x, y] in a Cartesian frame be *rotated* about the origin through an angle of

90°; let the new position be labeled P'[x', y']

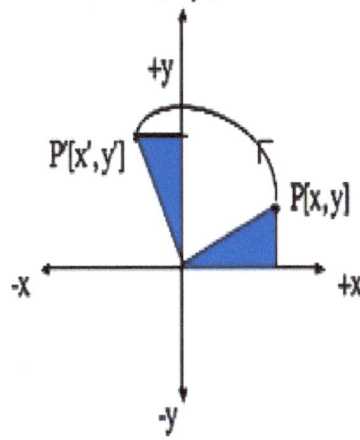

We see that the new coordinates are related to the old coordinates as follows:

$$x' \text{ (new)} = \quad -y \text{ (old)}$$

and

$$y' \text{ (new)} = +x \text{ (old)}$$

where we have written the x's and y's in different *columns*.

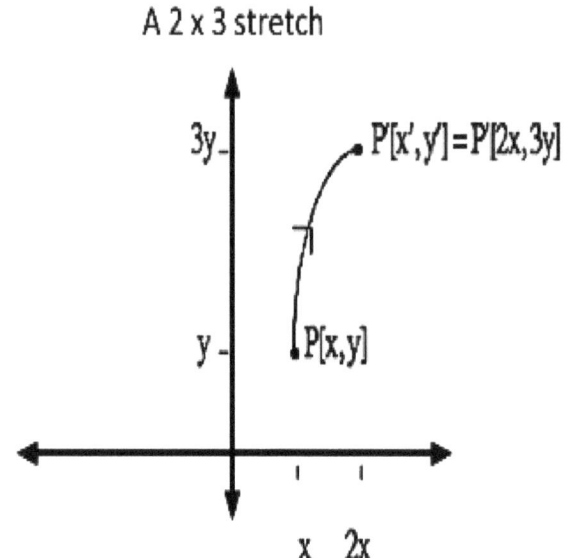

A 2 x 3 stretch

The new coordinates are related to the old coordinates by

$$x' = 2x$$

and

$$y' = 3y$$

In general, $x' = ax + by$
$$y' = cx + dy$$

In the 1840's, Cayley recognized the key role of the coefficients in characterizing the transformation of a coordinate pair [x, y] into the pair [x', y']. He therefore "separated them out", writing the pair of equations in column-form, thus:

$$x' = ax + by \qquad\qquad\qquad \longrightarrow \qquad \begin{pmatrix} x' \\ y' \end{pmatrix} = \begin{pmatrix} a & b \\ c & d \end{pmatrix} \begin{pmatrix} x \\ y \end{pmatrix}$$
$$y' = cx + dy$$

This is a *single* equation that represents the original *two* equations. We can write it in the symbolic form:

$$\mathbf{P' = MP}$$

The matrix M transforms the point P into the point P'

Geometry and Matrix Transformations

In 1874, Felix Klein introduced the "Erlangen program" in which he proposed that *all* branches of geometry can be described in terms of groups of matrix transformations. This was a radical departure from previous approaches. Euclidean geometry can be studied in terms of Cayley's rotation, translation and reflection matrices.

A proof of Pythagoras' Theorem

Invoking the invariance of length and angle under rotations and translations, the proof of Pythagoras' theorem – the most important theorem in Euclidean geometry – is immediate:

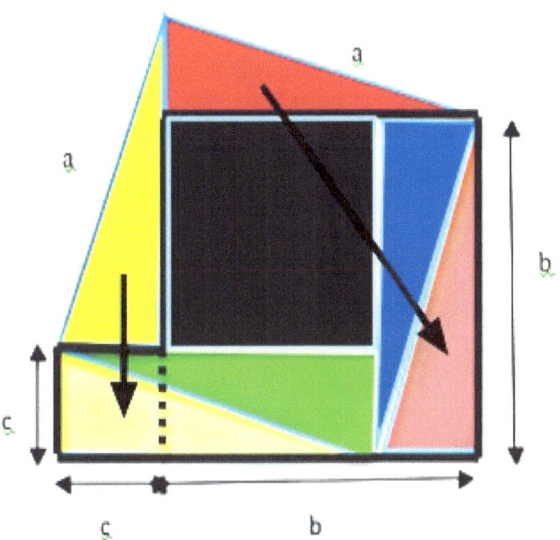

Moving the red triangle into the yellow, green and blue locations, the area of four triangles plus the area of the black square = a^2. Rearranging the total area by moving the red and yellow triangles, as shown, the new *conserved* area is the sum of the areas of two squares, $b^2 + c^2 = a^2$.

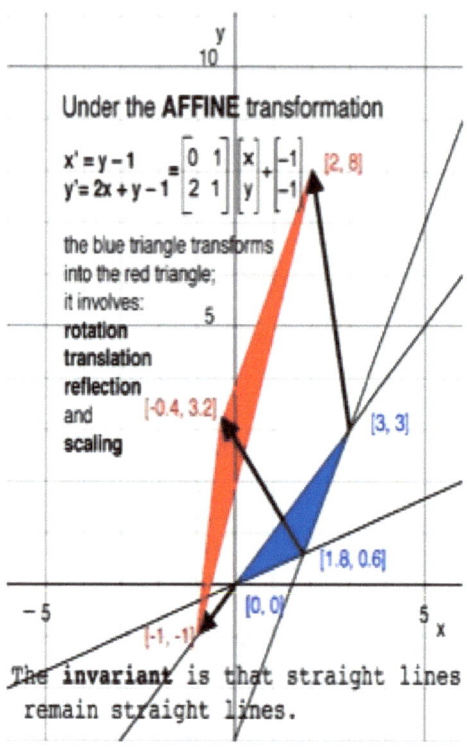

Under the **AFFINE** transformation

$$x' = y - 1$$
$$y' = 2x + y - 1 \quad = \begin{bmatrix} 0 & 1 \\ 2 & 1 \end{bmatrix} \begin{bmatrix} x \\ y \end{bmatrix} + \begin{bmatrix} -1 \\ -1 \end{bmatrix}$$

the blue triangle transforms
into the red triangle;
it involves:
rotation
translation
reflection
and
scaling

[2, 8]

[-0.4, 3.2]

[3, 3]

[1.8, 0.6]

[0, 0]

[-1, -1]

The **invariant** is that straight lines remain straight lines.

(In transforming straight lines, the ratio of distances is an invariant: e.g. the midpoint of a line remains the midpoint after the transformation).

All triangles are transformed into other triangles; this is a generalization of the concept of congruence, or similarity, in Euclidean geometry.
("affine" – "affinity with")

Operators and Complex Numbers

Let $R(90^0)$ be the operator that rotates the point p =[a, 0]
through 90^0 about the origin, giving the new point p'=[0, a].
If we rotate the point p' through 90^0 we obtain the point
p'' = [-a, 0].
Symbolically,

$$R^2(90^0)[a, 0] = [-a, 0] = -1[a, 0].$$

The operator $R(90^0)$ is equivalent to **√-1** (written i).
A number of the form

$$z = x + iy$$

is said to be "complex" ; x is the *real* part, and y is the
imaginary part of z.

82

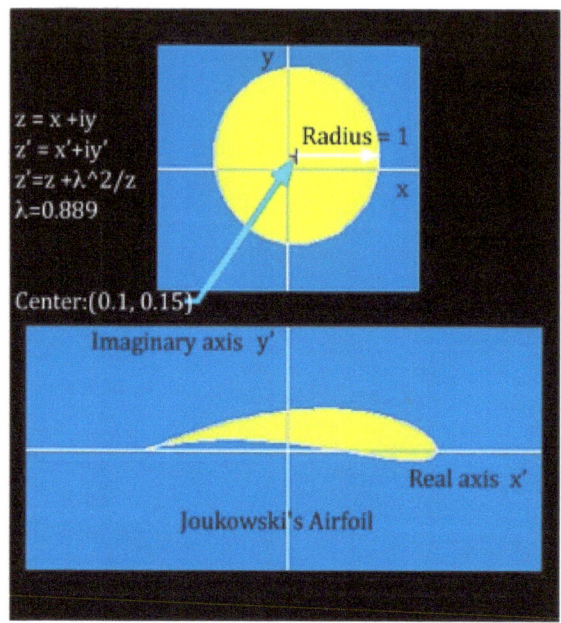

A *Joukowski* transformation in the complex plane. The circle transforms into an "airfoil" section; the transformation is used extensively in the design of aircraft. The well-understood pattern of airflow around the cylinder transforms exactly into the airflow pattern around the complicated airfoil section.

An example of the changes in the shape of a Joukowski airfoil that result from very small changes in the parameters of the transformation.

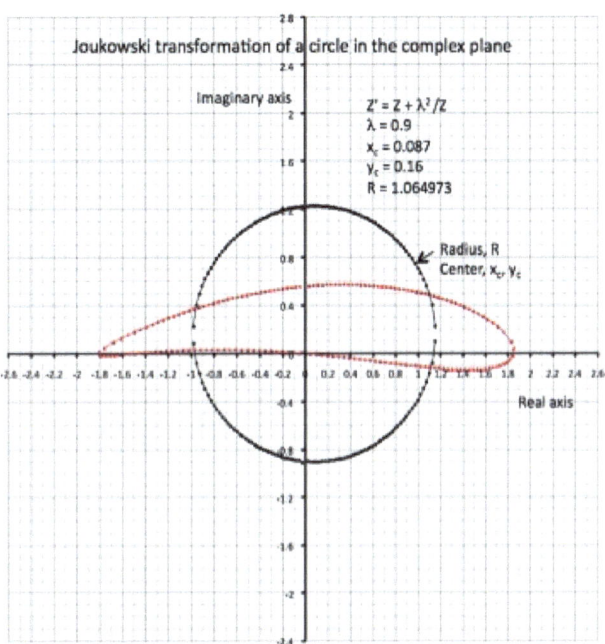

E = Mc²: a neo-Newtonian approach

April, 2014

Newton was well aware of the great difficulties that arise in any theory of the gravitational interaction between two masses not in direct contact with each other. In the *Principia*, he assumes, in the absence of any knowledge of the speed of propagation of the gravitational interaction, that the interaction takes place *instantaneously*. However, in letters to other luminaries of his day, he postulated an intervening agent between two separated masses — an agent that requires a *finite time* to react. In the early 19th century, the problem of understanding the interaction between spatially separated objects appeared in a new guise, this time in discussions of the electromagnetic interaction between *charged* objects. Faraday introduced the idea of a *field of force with dynamical properties*. In the Faraday model, an accelerating electric charge acts as the *source* of a dynamical electromagnetic field that travels at a finite speed through space-time, and interacts with a distant charge. Energy and momentum are thereby transferred from one charged object to another distant charged object.

Maxwell developed Faraday's idea into a mathematical theory — the electromagnetic theory of light — in which the speed of propagation of light appears as a fundamental constant of Nature. His theory involves the differential equations of motion of the electric and magnetic field vectors; the equations are not invariant under the Galilean transformation but they are invariant under the Lorentz transformation. (The discovery of the transformation that leaves Maxwell's equations invariant for all inertial observers was made by Lorentz in 1897). Einstein's Special Theory of Relativity is a theory in which there is but one universal constant, v_F say, for the speed of propagation of *any* massless dynamical field in a vacuum. In Einstein's original paper on Special Relativity he introduced the postulate that the speed of light in free space is a fundamental constant, usually written c. The universal constant v_F is not only the speed of light in free space but also the speed of the gravitational field, v_G say, in the void between interacting masses. Before discussing the notion of a gravitational field, a brief account of the essential work of Kepler is given.

In 1605, Johannes Kepler discerned two key properties of the orbital motions of the planets:

 1. Planets move in ellipses with the Sun at one focus.

 2. The radius vector sweeps out equal areas in equal   times.

These *planetary laws* demonstrate the crucial role that mathematical models play in the progress of science.

He arrived at these conclusions in *reverse order*. He first determined that the radius vector of a planet's "oval shaped" path sweeps out equal areas in equal times, and later found that the "ovals" were *ellipses*. The two laws are illustrated in the following diagram.

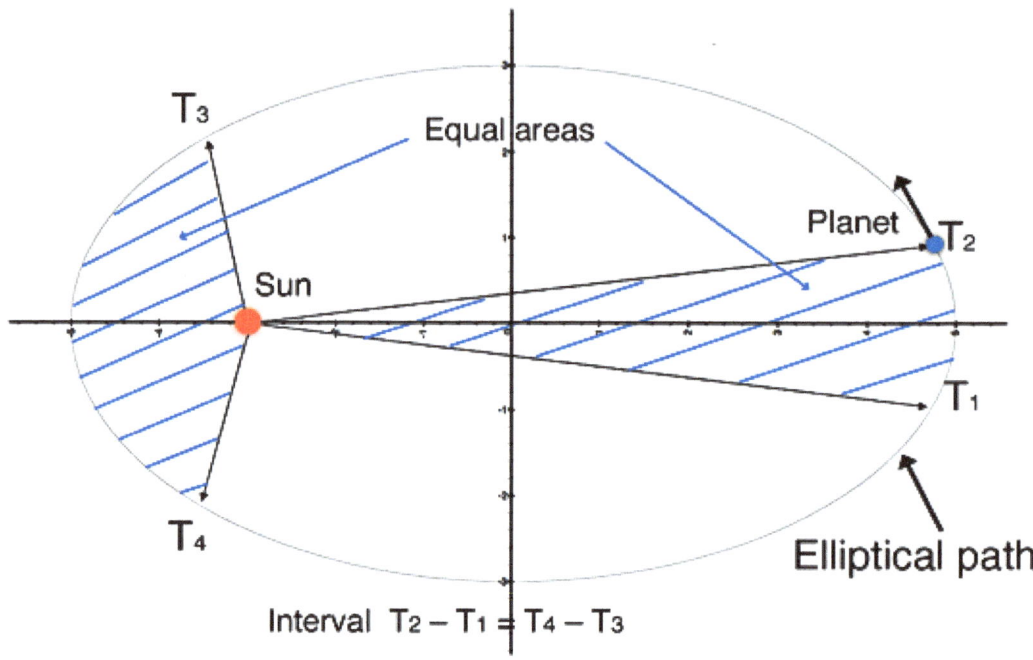

Figure 1. The first two laws of planetary motion

More than twelve years went by before Kepler discovered another law of planetary motion. He worked continuously during this period, searching for other possible regularities in planetary motion. The complexity of the astronomical data, coupled with the very large numbers involved in his calculations, made his task extremely difficult. By shear chance, a breakthrough in calculating and manipulating large numbers occurred in 1614 when John Napier, a Scot, introduced *logarithms.* Kepler learned of Napier's logarithmic method in late 1616. He used this new analytical tool, and discovered the following remarkable property of planetary motion:

"The proportion between the periodic times of any two planets is precisely one-and-a-half times the proportion of the mean distances."

In modern form his *third law* is usually written "the square of the orbital period of a planet is proportional to the cube of its semimajor axis".

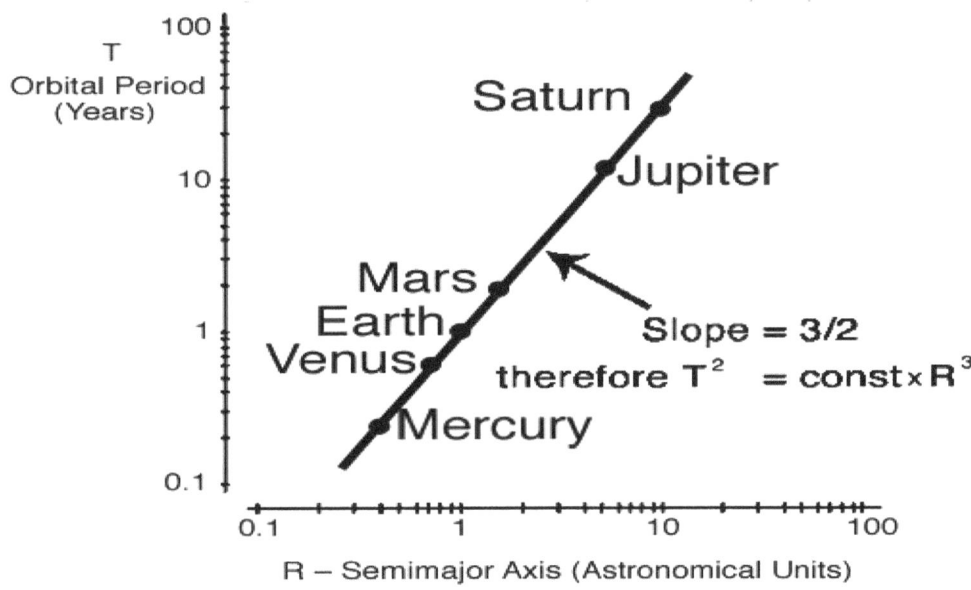

Figure 2. The slope of the (linear) relation between log(T) and log(R) is 3/2.

We now discuss another key discovery made by Newton, namely that *an object moving with constant speed in a circle is always accelerating towards the center of the circle.* We can understand this non-intuitive result by considering the following diagram that shows the "vectors" involved

in the motion. Vectors possess both *magnitude* and a sense of *direction*.
A particle of mass m at the position A has a velocity $\mathbf{v}_A$ tangential to the
circle at A. At a time Dt later, the particle is at position B and has a
velocity vector $\mathbf{v}_B$. The lengths of the vectors are the same.

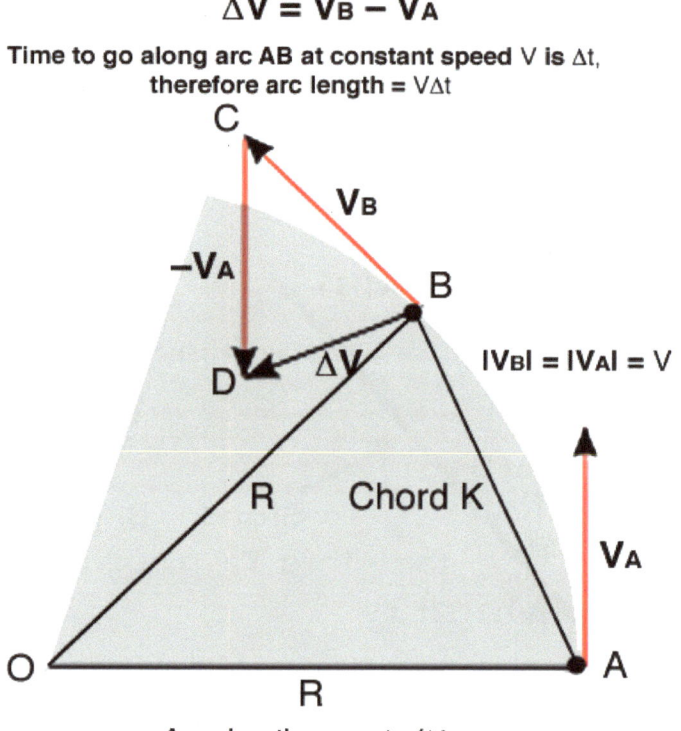

$$\Delta V = V_B - V_A$$

Time to go along arc AB at constant speed V is Δt,
therefore arc length $= V\Delta t$

$|V_B| = |V_A| = V$

Acceleration $\mathbf{a} = \Delta v / \Delta t$
As $\Delta t \to 0$, A $\to$ B and $\Delta \mathbf{v}$ points towards the center O,
therefore $\mathbf{a}$ points towards O.
The triangles ABO and BCD are similar, therefore $K/R = \Delta v/v$.
As $\Delta t \to 0$, K $\to$ AB $\to$ $V\Delta t$, therefore $a = \Delta v/\Delta t \to v^2/R$

Figure 3. Uniform motion in a circle: acceleration towards the center.

If v is the constant speed and R is the radius of the circle then the
magnitude of the acceleration is v^2/R.

Planetary motion in a circle and the inverse square law of gravitation

Consider the case of a planet of mass m moving with constant speed v in a circle of radius R about the Sun. Newton's 2nd law states that the force acting on a mass is directly related to its acceleration. The acceleration lasts while the force lasts. Newton therefore introduced a force **F** associated with the acceleration **a** of the planet towards the Sun, located at the center of the orbit:

$F = ma = mv^2/R$.

He knew that the motions of the planets obey Kepler's laws, and that the 3rd law gives a relation between the period T and the radius R of the orbit:

$T^2 = kR^3$ where k is a universal constant.

For a circular orbit the period T and the radius R are related to the constant speed v:

$v = 2\pi R/T$, and therefore $v^2/R = 4\pi^2 R/T^2$.

Substituting this result in the expression for F, and using $T^2 = kR^3$, we obtain

$F = mv^2/R = m4\pi^2 R/kR^3 = (m4\pi^2/k)/R^2$

$= Constant \cdot (1/R^2) \rightarrow$ **proportional to 1/R²**

Newton's 3rd law equates action and reaction; he therefore surmised that the *gravitational force* between the planet and Sun must also include the mass of the Sun, M_s; he therefore wrote

$F = GM_s m/R^2$

where G is a universal constant.

This is the first *"symmetry"* argument to be used in Physics. He went further, and replaced M_s and m by two masses m_1 and m_2 so that

$F = G\, m_1 m_2 /R^2$.

The "Mass" of the Gravitational Field

We can gain some insight into the dynamical properties associated with the interaction between distant masses by investigating the effect of a finite speed of propagation v_G, of the gravitational interaction, on Newton's Laws of Motion. Consider a non-orbiting mass M, at a distance R from a mass M_S, simply falling from rest with an acceleration **a**(R) towards M_S. According to Newton's Theory of Gravitation, the magnitude of the force on the mass M is

$$|\mathbf{F}(R)| = GM_S M/R^2 = Ma(R), \tag{1}$$

We therefore have

$$a(R) = GM_S/R^2. \tag{2}$$

(Cancelling the masses M in equation (1) assumes that "inertial" mass is equal to "gravitational" mass). Let Δt be the time that it takes for the *gravitational interaction* to travel the distance R at the speed v_G, so that

$$\Delta t = R/v_G. \tag{3}$$

In the time interval Δt, the mass M moves a distance ΔR towards the mass M_S;

$$\Delta R = a(R)\Delta t^2/2 \text{ (for "constant" } a(R))$$

$$= (GM_S/R^2)\Delta t^2/2$$

$$= (GM_S/R^2)(R/v_G)^2/2. \tag{4}$$

Consider the situation in which the mass M is in a circular orbit of radius R about the mass, M_S. Let $\mathbf{v}(t)$ be the velocity of the mass M at time t, and $\mathbf{v}(t + \Delta t)$ its velocity at $t + \Delta t$, where Δt is chosen to be the *interaction travel time*. Let us consider the motion of M if there were *no mass M_S present*, and therefore no interaction; the mass M then would continue its motion with constant velocity $\mathbf{v}(t)$ in a straight line (the first Law of Motion holds). We are interested in the difference in the positions of M at time $t + \Delta t$, *with and without* the mass M_S in place. We have, to a good approximation:

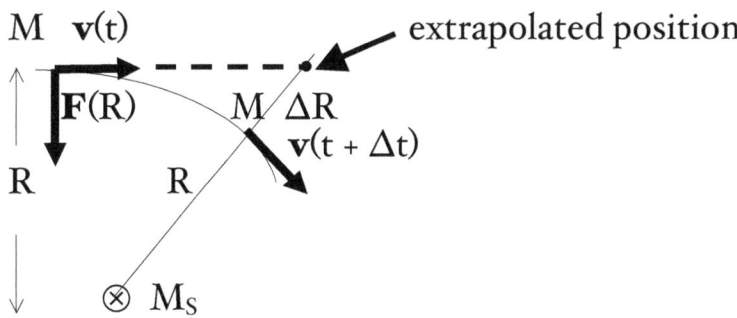

Figure 4. A model for estimating the gravitational field "mass"

The magnitude of the gravitational force, F_{EX}, at the extrapolated position, with M_S in place, is

$$F_{EX} = GM_SM/(R + \Delta R)^2 \tag{5}$$

$$= (GM_SM/R^2)(1 + \Delta R/R)^{-2}$$

$$\approx (GM_SM/R^2)(1 - 2\Delta R/R), \text{ for } \Delta R << R. \tag{6}$$

(The approximation in (6) is an application of Newton's *binomial theorem*). Substituting the value of ΔR obtained above, we find

$$F_{EX} \approx GM_SM/R^2 - (GM_SM/Rv_G^2)(GM_S/R^2). \tag{7}$$

Newton's 3rd Law states that

$$\mathbf{F}_{MS, M} = -\mathbf{F}_{M, MS} \tag{8}$$

This Law is true, however, for *contact* interactions only. For all interactions that take place between separated objects, there is a mismatch between the action and the reaction. *It takes time for one object to respond to the presence of the other.*

In the present example, we obtain a good estimate of the mismatch by taking the difference between $F_{EX}(R + \Delta R)$ and $F(R)$, namely

$$F_{EX}(R + \Delta R) - F(R) \approx (GM_SM/Rv_G^2)(GM_S/R^2). \tag{9}$$

On the right-hand side of this equation, we note that the term (GM_S/R^2) has dimensions of "acceleration", and therefore the term (GM_SM/Rv_G^2) must have dimensions of "mass". This term is an estimate of the "mass" associated with the interaction, itself. *The space between the interacting masses must be endowed with this effective mass if Newton's 3rd Law is to include non-contact interactions,* $M_{GRAV} = GM_S M/Rv_G^2$. The appearance of the term v_G^2 in the denominator of this effective mass term has a special significance. The term $GM_S M/R$ has dimensions of "energy" and therefore we introduce an effective energy $E_{GRAV} = GM_S M/R$ associated with the gravitational interaction:

$$E_{GRAV} = M_{GRAV} \bullet v_G^2 = GM_S M/R, \text{ a neo-Newtonian equation} \quad (10)$$

*This is the "energy stored in the gravitational field" between the two interacting masses. Note that it has a 1/R-dependence — the correct form for the **potential energy** associated with a 1/R² gravitational force.*

The notion of a dynamical field of force is a necessary consequence of the finite propagation time of the interaction.

In Einsteinian Relativity, the speed of propagation of the gravitational field, v_G, is equal to the speed of light, c; equation (10) can therefore be written

$$E_{GRAV} = M_{GRAV} \bullet v_G^2 = M_{GRAV} \bullet c^2 \quad (11)$$

Although this equation applies to the gravitational field, Einstein introduced the *general form*

$$E = mc^2 \quad (12)$$

It should be noted that the speed of propagation of the gravitational field has not been *measured*; it is a *theoretical prediction*.

The general concept of energy-mass equivalence

In Newton's *Optiks* (1717) he asks, in Query 30:

"Are not the gross bodies and light convertible into one another, and may not bodies receive much of their activity from the particles of light which enter their composition?"

Following Maxwell's acclaimed *Theory of Electromagnetism* (1865) and his concept of *radiation pressure* (1871), many physicists and mathematicians discussed the relationship between the energy, momentum, and mass associated with the electromagnetic field. They included Poynting (1874), Thomson (1881), Heaviside (1888), Searle (1897), Wien (1900), Poincare (1900, 1905), Abraham (1902), Hasenoehrl (1904), Lorentz (1904), and Einstein (1905).

Einstein's argument was based on his *Theory of Relativity* and therefore it implied the *universality of the equivalence*.

An example: the lifetime of the Sun

In 1904, Rutherford proposed that the Sun's energy output could be maintained by an internal source of *radioactivity*. (His proposal was made only a few years after the discovery of radioactivity).

It is now known that the Sun's energy originates in *nuclear fusion* reactions in its interior where the temperatures and pressures are so enormous that these reactions are possible. In such processes, mass is converted into energy in the form of high-energy gamma rays and elementary particles. The gamma rays lose energy on their way to the surface, and emerge as low-energy electromagnetic radiation.

In the free state, the neutron, a fundamental constituent of all complex nuclei, is unstable. Its lifetime is about 10 minutes. It transforms into a proton, an electron, and an entity called an anti-neutrino:

$$n^\circ \to p^+ + e^- + \bar{\nu}^\circ \tag{13}$$

This process can be transformed as follows (it can only occur in the presence of a nuclear force field because the proton mass is less than the neutron mass):

$$p^+ \to n^\circ + e^+ + \nu^\circ, \tag{14}$$

proton → neutron + positron + neutrino.

At temperatures greater than 10 million degrees two protons (the nuclei of hydrogen atoms) have sufficient energy to overcome the large Coulomb repulsive force between them, and they can "fuse". For stars with masses about equal to that of our Sun, and slightly less, the predominant nuclear fusion reaction is known as the *proton-proton cycle:*

If one of a pair of protons transforms into a neutron, plus positron, plus neutrino, the other can combine with the neutron to form a complex nucleus called a deuteron, d

$$p + n \to d. \tag{15}$$

The deuteron now combines with another proton to form helium-3:

$$p + d \rightarrow He^3 + gamma\ ray. \tag{16}$$

If two more protons combine to form another He^3 nucleus then the two He^3 nuclei can fuse to form He^4 plus two protons:

$$2He^3 \rightarrow He^4 + 2p. \tag{17}$$

The mass of the final He^4 nucleus is less than the mass of the four original protons by 0.7%. This "missing mass" transforms into energy according to the equation $E = mc^2$. The fusion process is illustrated:

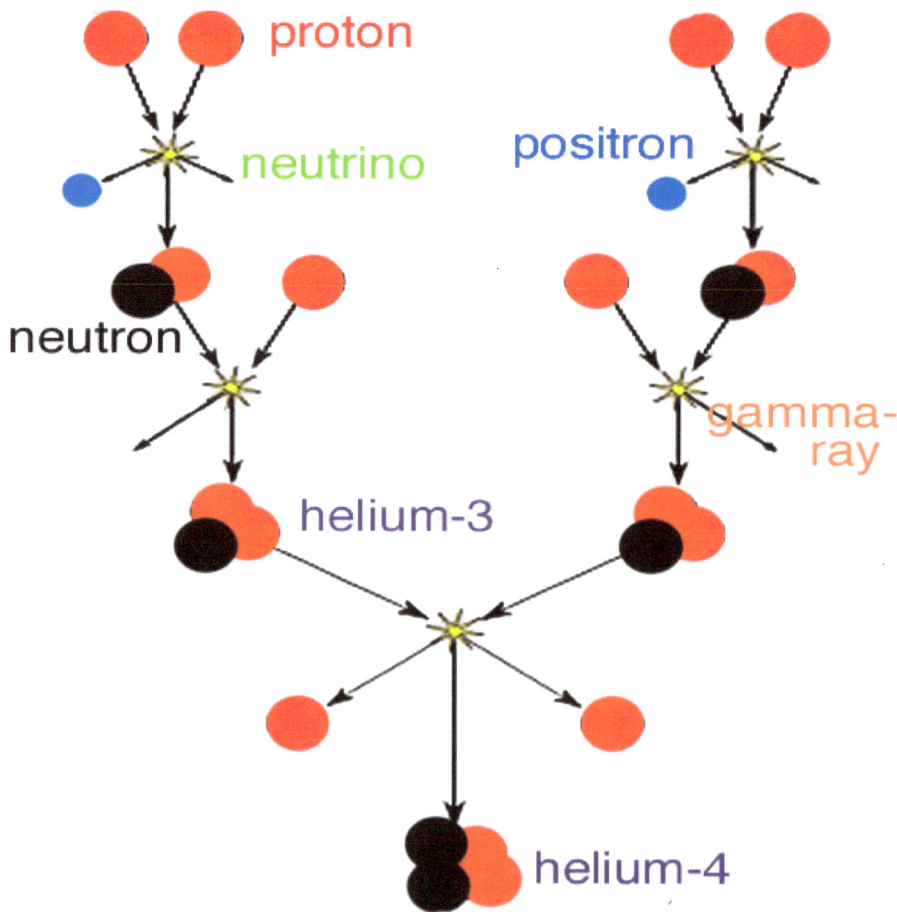

Figure 5. The double proton-proton fusion chain reaction cycle. The final He^4 nucleus has 0.7% less mass than the original four protons.

The mass of the Sun is about 2×10^{30} kilograms (333,000 times the mass of the Earth). Measurements show that the total radiation from the surface of the Sun is about 4×10^{26} watts. The available energy from the Sun is

$$E = 0.007 \times M_{available} \times c^2 \tag{18}$$

where $M_{available}$ is the mass of the Sun that can undergo nuclear fusion reactions. Only the central part of the Sun has the very high temperatures needed to produce fusion reactions. A detailed model of the Sun is required to estimate this mass; a reasonable model indicates that $M_{available}$ is about 10% of the total mass, therefore

$$E = 0.007 \times 0.1 \times M_{Sun} \times c^2 \tag{19}$$

The total energy available is therefore

$$E = 0.0007 \times 2 \times 10^{30} \times 9 \times 10^{16} \text{ Joules (the unit of energy in the}$$
$$\text{SI system)}$$

$$\approx 1.3 \times 10^{44} \text{ Joules.} \tag{20}$$

The lifetime of the Sun (in seconds!) is therefore

$$T_{Sun} = 1.3 \times 10^{44} \text{ Joules} / 4 \times 10^{26} \text{ (Joules per second (= watts))}$$

$$\approx 10 \text{ billion years.} \tag{21}$$

The Sun is now about 5 billion years old, and therefore in another 5 billion years it will have lost all its mass capable of generating fusion reactions. It will become a Red Giant, engulfing much of the solar system, then a White Dwarf, and eventually a Black Dwarf, providing material that will drift through the void, perhaps forming new worlds.

Appendix

Einstein's Theory: relativistic mass, momentum and energy

In *geometry*, the scalar product of two vectors **A** with components $[a_1, a_2]$ and **B** with components $[b_1, b_2]$ is important because it is *invariant* under the operations of rotation and translation of the coordinate system in which the components are measured. Explicitly,

$\mathbf{A} \cdot \mathbf{B} = a_1 b_1 + a_2 b_2 = \mathbf{A}' \cdot \mathbf{B}' = a_1' b_1' + a_2' b_2'$

where, for the **A**- vector

$$\begin{pmatrix} a_1' \\ a_2' \end{pmatrix} = \begin{pmatrix} \cos\theta & \sin\theta \\ -\sin\theta & \cos\theta \end{pmatrix} \begin{pmatrix} a_1 \\ a_2 \end{pmatrix}$$

in which θ is the angle of rotation of the coordinate system (from *unprimed* to the *primed* system).

In *space-time*, Nature prescribes the difference-of-squares as *invariant* under the Lorentz transformation that relates measurements in one inertial frame to measurements in another. For two events $E_1[ct, x]$ and $E_2[ct, -x]$, their scalar product is

$\quad E_1 \cdot E_2 = (ct)^2 - x^2$

where we have chosen the direction of E_2 to be opposite that of E_1, thereby providing the necessary negative sign in Nature's invariant.

The Lorentz transformation relates an event $[ct, x]$ in one inertial frame to the same event $[ct', x']$, measured in another inertial frame moving at constant relative speed v_x, as follows

$$\begin{pmatrix} ct' \\ x' \end{pmatrix} = \begin{pmatrix} \gamma & -\beta\gamma \\ -\beta\gamma & \gamma \end{pmatrix} \begin{pmatrix} ct \\ x \end{pmatrix}$$

where $\gamma = (1 - \beta^2)^{-1/2}$ and $\beta = v_x/c$.

In terms of finite differences of time and distance, we obtain

$$(c\Delta t)^2 - (\Delta x)^2 = (c\Delta\tau)^2 = \text{invariant}$$

where $\Delta\tau$ is the *proper time interval*. It is related to Δt by the equation

$$\Delta t = \gamma\Delta\tau.$$

In Newtonian mechanics, the quantity momentum, the product of the mass of an object and its velocity, plays a key role. In Einsteinian Mechanics, velocity, mass, momentum, and kinetic energy are redefined. These fundamental redefinitions are a direct consequence of the replacement of Newton's absolute time interval Δt_N by Einstein's velocity-dependent interval:

$$\Delta t_E = \gamma\Delta\tau.$$

The Newtonian momentum $\mathbf{p}_N = m_N\mathbf{v}_N = m_N\Delta\mathbf{x}/\Delta t_N$ is replaced by the Einsteinian momentum

$$\mathbf{p}_E^+ = m_o\mathbf{v}_E = m_o\Delta[ct, \mathbf{x}]/\Delta\tau$$

where m_o is the *rest mass*. The reason for introducing the + upper index will become clear, later.

The momentum can be written

$$\mathbf{p}_E^+ = m_o[c\Delta t/\Delta\tau, \Delta\mathbf{x}/\Delta\tau]$$

$$= m_o[\gamma c, (\Delta\mathbf{x}/\Delta t)(\Delta t/\Delta\tau)]$$

$$= m_o[\gamma c, \gamma\mathbf{v}_N].$$

We introduce the momentum vector in which the direction of the x-component is reversed, giving

$$\mathbf{p}_E^- = m_o[\gamma c, -\gamma\mathbf{v}_N].$$

Forming the scalar product, we obtain

$$\mathbf{p}_E^+ \cdot \mathbf{p}_E^- = m_o^2(\gamma^2 c^2 - \gamma^2\mathbf{v}_N^2) \text{ (a Lorentz-invariant form)}$$
$$= m_o^2 c^2,$$

because $\mathbf{v}_E^+ \cdot \mathbf{v}_E^- = c^2$.

Multiplying throughout by c^2, and rearranging, we find

$$m_o^2 c^4 = \gamma^2 m_o^2 c^4 - \gamma^2 m_o^2 c^2\mathbf{v}_N^2.$$

Now, γ is a number, and therefore γ multiplied by the rest mass m_o also is a *mass*; let us therefore denote it by the symbol m:

$m = \gamma m_o$, called the relativistic mass. It is velocity-dependent.

We can then write

$$(m_o c^2)^2 = (mc^2)^2 - (c\mathbf{p}_E)^2.$$

The quantities mc^2 and $m_o c^2$ have dimensions of *energy*; we therefore denote them by the symbols E and E_o, respectively, so that

E = mc^2, Einstein's renowned equation,

and

$E_o = m_o c^2$, the *rest energy* of the mass.

We therefore obtain the *fundamental invariant* of relativistic dynamics:

$$E_o^2 = E^2 - (\mathbf{p}_E c)^2 = E'^2 - (\mathbf{p}_E'c)^2 \text{ in any other inertial (prime)-frame.}$$

This invariant includes entities with zero rest mass. For example, a photon of total energy E_{PH}, and momentum $\mathbf{p}_{PH}$ satisfies the equation

$$o = E_{PH}^2 - (p_{PH}c)^2, [\text{note, } \mathbf{p}_{PH} \cdot \mathbf{p}_{PH} = p_{PH}^2]$$

and therefore

$E_{PH} = p_{PH}c.$

No violations of Einstein's Theory of Special Relativity have been found in any tests of the theory, carried out to this day.

Some Drawings, Paintings and Graphics
Exhibited at the Koerner Center for Emeritus Faculty,
Yale University, April 2004

The following selection of drawings, paintings and graphics were exhibited at the Koerner Center in 2004.

A West Country image painted in late 1942 during my first year at the Coopers' Company's School. I was just 12 years old at the time. The school had been evacuated from Bow in the East End of London to the relative safety of Frome in Somerset.

In early 1945, I cycled from Frome to Wells, the smallest City in England and I made a detailed drawing of the magnificent cathedral. On my return to my home in London, at the end of The War, I completed the work using watercolors.

In 1946, I came across a small black and white image of Cotman's watercolor of Greta Bridge, and I painted my version of it.

In 1981, I went on a pilgrimage to Newton's birthplace, sat under a descendant of the renowned apple tree, and did this sketch of Woolsthorpe Manor.

In 1964, we were living in Newbury, Berkshire. At that time, I did a detailed drawing of Newbury Bridge. After arriving in the US in 1965, I did several paintings based on my original sketch. This example was painted in 2004.

A graphical image based on the photonuclear cross sections for the interaction of high-energy gamma rays with oxygen-16 and carbon-12 nuclei. I made the measurements in 1962.

A graphical image of a hedgerow (1978).

A watercolor of the High Sierras (2000/2004 version).

Hopiland: a watercolor (1961/2004 version).